Unlocking ChatGPT for Beginners

A Guide to Master Prompt Engineering, Engage in Intelligent Conversations, Improve Efficiency, Decision Making and Achieve Breakthrough Results

C. P. Bailey

Contents

Introduction

"The more you know, the more you realize how much you don't know."
Aristotle

The Four Stages Of Competence

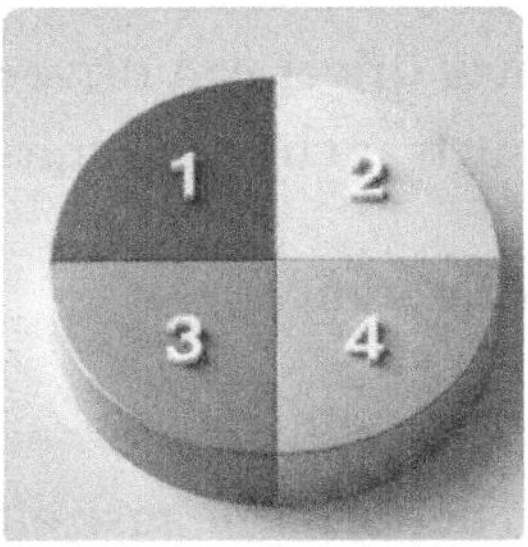

The **Four Stages of Competence** model, also known as the **Conscious Competence Ladder**, is attributed to **Noel Burch**, who developed it in the **1970s** while working for **Gordon Training International.** This model was initially created to help trainers understand and guide the process of learning and skill acquisition.

The **four states of competency** describe the stages individuals go through as they acquire and master new skills or knowledge. The stages are often depicted as a learning framework. Here are the four stages:

1. **Unconscious Incompetence**
    - **Definition:** The individual is unaware of their lack of skill or knowledge. They don't know what they don't know

   - **Example:** Someone who doesn't know how to drive but isn't aware of the complexities involved in learning to drive.

2. **Conscious Incompetence**

   - **Definition:** The individual becomes aware of their lack of knowledge or skill. They know they don’t know something and recognize the gap.

   - **Example:** A person starts driving lessons and realizes how much there is to learn.

3. **Conscious Competence**

   - **Definition:** The individual can perform the skill but must think and concentrate to execute it correctly.

   - **Example:** A new driver can drive a car but must focus on every action, such as checking mirrors and shifting gears.

4. **Unconscious Competence**

   - **Definition:** The skill becomes second nature. The individual can perform the task effortlessly without much conscious thought.

   - **Example:** An experienced driver drives intuitively, without actively thinking about the mechanics of driving.

This framework is widely used to understand and facilitate skill acquisition in educational and professional development contexts. We will utilize the **Four Stages** to represent our progress through our journey into unlocking ChatGPT.

Challenge yourself as you move through each chapter and ask yourself where you are in the **Four stages of learning and skill acquisition**. For those who wish to dive deeper into how Four Stages and AI training can benefit almost every industry, see the Bonus Content section of this book.

All right, let's jump right into our journey to unlock ChatGPT.

Let's get Started

A few months ago, I found myself sitting at my desk, surrounded by stacks of papers and endless to-do lists. I felt stuck and overwhelmed by a project that needed a fresh perspective. That's when I turned to ChatGPT. I decided to ask it a seemingly simple question: "How can I approach this problem differently?" The response was unexpected and eye-opening. ChatGPT suggested a novel angle I hadn't considered. This small interaction sparked a breakthrough, moving my project forward in ways I hadn't imagined. It was a moment that changed how I viewed AI—no longer as just a tool but as a partner in creativity and problem-solving.

Many adults today find themselves in a similar position when engaging with AI. The potential is there, but the path isn't always clear. It's easy to feel overwhelmed by technology or unsure about how to ask the right questions. This challenge is more common in our tech-driven world than you might think. People often struggle with formulating effective questions or feel intimidated by the complexity of AI. This book addresses these common pain points by offering a straightforward, approachable guide to navigating the world of ChatGPT.

The promise of this book is simple yet powerful: to empower you to unlock the potential of ChatGPT. You'll learn how to enhance your career development, boost productivity, and foster personal growth. This isn't just another resource on AI. It's a practical guide designed to offer unique insights and techniques you won't find elsewhere.

What sets this book apart is its focus on practical applications. We'll explore how to craft effective prompts that lead to better AI responses. You'll discover actionable strategies that can be applied to real-life challenges. Whether you want to improve communication, solve complex problems, or get a fresh perspective, this book has something to offer.

Throughout these pages, we'll cover key topics such as prompt engineering and continuous learning. You'll learn how to overcome common obstacles when interacting with AI. By the end of this book, you'll gain improved communication skills and increased tech literacy and be ready to apply these skills in your daily life.

Allow me to introduce myself. I am passionate about helping adults engage with AI in meaningful ways. My experience in this field has equipped me with the insights needed to guide you through this journey. Having worked with various AI technologies, I understand their challenges and opportunities. My goal is to make AI accessible and beneficial for everyone, regardless of their background.

You can expect a deeper understanding of ChatGPT as you read this book. You'll gain practical skills for engaging with AI and the confidence to apply these skills in various aspects of your life. By the end of this journey, you'll be equipped to unlock the full potential of ChatGPT and transform your approach to AI interactions.

I encourage you to engage actively with the content and remain curious. Experiment with ChatGPT and explore its capabilities beyond the examples provided. Curiosity is your greatest ally in this learning process, and the more you engage, the more you'll discover.

So, are you ready to unlock ChatGPT's potential? Join me on this journey and transform how you interact with AI. Let's dive in and explore the possibilities together.

Chapter One

Foundations of ChatGPT

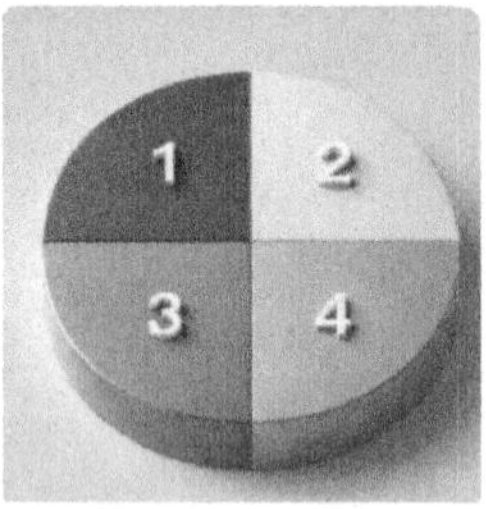

Have you ever been stumped by a question, only to find the answer popping out during a casual chat with a friend? That's the magic of conversation, where ideas flow freely and solutions emerge almost effortlessly. Now, imagine having that kind of enlightening dialogue with a machine. That's what ChatGPT offers—a chance to engage in a natural conversation with an AI that can help untangle complex problems, inspire creativity, and enhance productivity. This chapter will introduce you to the foundational aspects of ChatGPT, offering you a clear understanding of what it is, how it functions, and how it can become a powerful tool in your daily life.

1.1 Understanding ChatGPT: What It Is and How It Works

At its core, ChatGPT is an AI language model crafted to engage in natural dialogue. It's designed to mimic human conversation, making it a valuable tool for everything from brainstorming ideas to providing customer support. The beauty of ChatGPT lies in its ability to handle a wide range of conversational applications, whether you're using it to draft emails, create content, or even solve a tricky math problem. It enhances communication efficiency by processing and generating human-like responses, allowing you to interact with technology intuitively and seamlessly.

The technology powering ChatGPT is both intricate and fascinating. It's built on a neural network architecture known as the transformer model. This model uses self-attention mechanisms, which means it can focus on different parts of a sentence as it processes information, much like how you might emphasize certain words or phrases when speaking. ChatGPT consists of components like an encoder and a decoder, which help it understand input and generate responses. The language model head predicts the next word in a sequence, while the dialogue generation head is fine-tuned to produce conversational replies. It's this combination that makes ChatGPT so adept at understanding and participating in a dialogue that feels genuine and meaningful.

When you first interact with ChatGPT, the experience begins with a simple setup. After signing up, you're greeted by an intuitive user interface that guides you through the process of initiating conversations. Whether you're using a free or paid version, the interface allows you to input your queries and receive responses in real-time. The cycle is straightforward: you type a question or prompt, and ChatGPT processes this input to produce a thoughtful response. This interaction is the heart of your engagement with ChatGPT, and it's all about asking the right questions to get the most relevant answers.

The effectiveness of ChatGPT hinges greatly on the quality of your input. Crafting clear and concise questions can significantly influence the AI's responses. Context and specificity are your allies here. For instance, asking, "What's the best way to improve my productivity?" is a good start. However, adding details such as "in a remote work setting" gives the AI the context to tailor its advice more closely to your situation. The better you articulate your query, the more insightful the AI's output will be, turning a simple interaction into a valuable exchange of ideas.

1.2 The Evolution of Conversational AI

The journey to what we now recognize as conversational AI is a fascinating tale of innovation and relentless curiosity. It all began in the 1960s with a program called Eliza, developed by Joseph Weizenbaum. Eliza was rudimentary by today's standards, using simple keyword recognition to simulate a conversation with a therapist. Despite its simplicity, it marked a pivotal moment in AI history, demonstrating the potential for machines to engage in human-like dialogue. Over the years, we've seen significant milestones, each contributing to our current understanding of AI. The transition from rule-based systems, which relied heavily on pre-defined scripts, to learning systems that adapt and evolve represents a quantum leap in technological advancement. This shift was fueled by breakthroughs in natural language processing (NLP) and machine learning, which have expanded the boundaries of what AI can achieve.

Central to these advancements is the development of NLP, a branch of AI focused on enabling computers to understand and generate human language. NLP has facilitated large-scale data analytics, improved customer satisfaction, and enhanced marketing insights. Its applications range from search engines that predict user intent to voice-activated assistants that respond to spoken queries. The integration of machine learning, which allows systems to learn from past interactions and improve over time, has further propelled conversational AI capabilities. These technologies have evolved hand-in-hand, creating more efficient, intuitive, and contextually aware systems. Machine learning algorithms have become sophisticated enough to provide personalized experiences, offering a glimpse into the future potential of AI.

In the early days, chatbots like Eliza offered limited interaction, constrained by their inability to process context or understand nuance. They operated on a set of predetermined rules, often producing responses that felt mechanical and disconnected from the flow of conversation. Fast forward to today, and the landscape is vastly different. Modern AI systems, like ChatGPT, have overcome many of these limitations. They can engage in complex, multi-turn conversations and respond with contextual awareness, offering a more human-like interaction. This progress is a testament to the remarkable advances in AI technology and the persistent efforts of researchers and developers who have pushed the boundaries of what machines can do.

Looking ahead, the future of conversational AI is full of promise and potential. Ongoing research continues to explore new frontiers, from enhancing AI's ability to understand emotional cues to developing systems that can learn with minimal human intervention. Emerging areas such as unsupervised learning and reinforcement learning offer exciting possibilities for creating even more advanced AI. The societal impacts are profound, with AI poised to revolutionize industries, improve quality of life, and address complex global challenges. As these technologies evolve, we find ourselves at the brink of a new era where conversational AI could become as ubiquitous as the internet, transforming how we interact with machines and each other in ways we are just beginning to imagine.

1.3 Key Components of AI: NLP and Machine Learning

Imagine telling your computer what you want, and it understands you perfectly—not just the words, but the sentiment behind them. This is where Natural Language Processing, or NLP, comes into play. NLP is the branch of AI that helps machines make sense of human language. It lets your smartphone's voice assistant understand your request to play your favorite song or what allows your email app to predict what you're typing next. At the heart of NLP are language understanding models, which parse the intricacies of grammar, context, and meaning to interpret what we say or type. They act almost like a translator, converting human language into data that a computer can process and respond to. Sentiment analysis is another fascinating aspect of NLP. It's like a mood ring for text, analyzing words to determine the emotional tone behind them. Businesses use it to gauge customer satisfaction, while social media platforms might use it to monitor public sentiment on trending topics. This capability helps companies understand how customers feel about their products or services in real-time.

Now, let's talk about machine learning, the secret sauce that powers much of today's AI. Machine learning is the process through which computers learn from data—lots of it. Imagine teaching a child to recognize different animals through pictures. Initially, they might guess wrong, but with more examples and corrections, they start getting better. Machine learning works similarly, using algorithms to sift through large datasets to find patterns and improve over time. This learning process involves training data, essentially a set of examples from which the AI learns. Over time, the system refines its accuracy through model optimization techniques, tweaking its parameters to enhance performance. This ongoing learning process is what allows AI to get smarter and more efficient, making it better at tasks like predicting the weather, recommending movies, or even diagnosing diseases.

The magic happens when NLP and machine learning come together, as they do in ChatGPT. These technologies work in tandem to power this advanced AI system. ChatGPT uses NLP to understand and generate responses in human language, while machine learning helps it improve its understanding and responses over time. The interaction between NLP and deep learning—the subset of machine learning that uses neural networks—allows ChatGPT to process language with remarkable nuance and depth. Transformer models, a type of deep learning architecture, play a crucial role here. They enable ChatGPT to process entire sentences simultaneously, understanding context and relationships between words in a way that older models couldn't.

This ability makes ChatGPT capable of holding conversations that feel natural and coherent.

Beyond ChatGPT, NLP and machine learning have a wide array of applications. Automated translation services, for instance, rely on these technologies to convert text from one language to another, breaking down communication barriers across the globe. Services like Google Translate use complex machine-learning algorithms to continually improve the accuracy and fluency of translations, making them indispensable tools for travelers and businesses alike. Speech recognition systems are another area where these technologies shine. They convert spoken language into text, allowing users to interact with their devices hands-free. Consider the convenience of dictating a text message while driving or asking your smart speaker to adjust the thermostat. These systems use NLP to understand the spoken words and machine learning to improve their accuracy over time, learning from countless interactions to better recognize accents, dialects, and speech patterns.

1.4 Navigating the User Interface: Practical Tips

Navigating the ChatGPT interface is like learning to drive a new car—it might seem complex at first, but once you get the hang of it, everything starts to feel intuitive. When you first log into ChatGPT, you'll be greeted by a clean and straightforward dashboard. This is your command center, where you'll spend most of your time. On the dashboard, you'll find features designed to make your experience smooth and efficient. There are options for starting new conversations, accessing recent chats, and customizing your

settings to suit your preferences. Customization options are quite handy. You can tweak the interface to match your style, such as adjusting the theme from light to dark mode to reduce eye strain or setting preferences that help the AI understand you better. These small, personal touches can make a big difference in how comfortable and engaging your experience becomes.

Getting comfortable with the interface is crucial because it dramatically enhances your user experience. When you know your way around, you can focus on the content of your conversations without being distracted by figuring out where to click next.

Efficient navigation techniques can save you time and frustration. For instance, using the search function to quickly locate previous interactions can be a lifesaver when revisiting past discussions. Moreover, accessing help and support features is straightforward. Whether through an FAQ section or a chatbot assistant, support is always a click away, ready to guide you through any hiccups you might encounter. Knowing how to find help quickly keeps your momentum going and reduces downtime, leaving you more time to engage with the AI.

Even with a user-friendly interface, everyone encounters technical issues from time to time. The good news is that most common problems have simple solutions. For example, if you experience connectivity issues, checking your internet connection or restarting your device often resolves the problem. Input errors can also occur, especially if you're typing quickly. Double-check your queries for typos or incomplete sentences before hitting enter. If you find that the AI isn't responding as expected, refreshing the page or logging out and back in can reset the system and clear any glitches. Having these troubleshooting tips at your fingertips can help you stay on track, minimizing disruptions to your workflow.

To elevate your efficiency even further, familiarize yourself with the shortcuts and tools that ChatGPT offers. Keyboard shortcuts can significantly speed up your tasks, much like knowing the shortcuts on your computer. For instance, using a shortcut to start a new conversation or switch between chats saves you time navigating through menus. Automated response tools can also boost productivity. They allow you to set up canned responses for frequently asked questions or repetitive tasks, letting the AI handle routine queries while you focus on more complex interactions. These features streamline your use of ChatGPT and enhance your overall productivity, enabling you to get more done in less time.

Interactive Element: Quick Start Checklist

1. **Explore the Dashboard:** Familiarize yourself with the layout and features.
2. **Customize Your Settings:** Adjust themes and preferences to enhance comfort.
3. **Learn Navigation Shortcuts:** Practice using search functions and keyboard shortcuts.
4. **Access Help Easily:** Know where to find FAQs and support features.
5. **Troubleshoot Common Issues:** Keep solutions for connectivity and input errors handy.
6. **Utilize Automated Tools:** Set up canned responses and explore productivity features.

By following this checklist, you can become adept at using ChatGPT, making every interaction smooth and productive.

1.5 AI Capabilities and Limitations: Setting Realistic Expectations

Imagine having a tool at your fingertips that can whip up creative content, whether it's for a blog post, a poem, or even a catchy slogan for your business. ChatGPT has become a go-to for many seeking a spark of creativity. It excels in generating content that feels fresh and engaging, offering suggestions that can inspire new ideas or provide a different perspective on a familiar topic. ChatGPT acts like a helpful assistant in customer service, ready to step in and easily handle inquiries. It can quickly produce responses to common questions, helping businesses manage customer interactions more efficiently.

This capability enhances productivity and allows human agents to focus on more complex issues that require a personal touch.

However, as remarkable as these capabilities are, it's essential to approach ChatGPT with a clear understanding of its limitations. One area where it may stumble is handling ambiguous queries. If a question is too vague or lacks context, ChatGPT might return a response that feels off-target or incomplete. It's not that the AI isn't capable, but rather that it relies heavily on the details provided to craft its answers. In scenarios where highly specialized information is required, ChatGPT might not always deliver the depth of expertise a human specialist can offer. For instance, while it can provide general medical advice or legal information, it shouldn't replace professional consultation in these fields. Recognizing these constraints is key to using ChatGPT effectively and ensuring that it complements rather than replaces human judgment and expertise.

To maximize what ChatGPT can offer, consider tailoring your prompts with clarity and precision. Think of it like giving directions to someone visiting your city for the first time—the more detailed you are, the easier it is for them to get where they need to go. By providing specific context or framing your questions clearly, you help guide ChatGPT in delivering more relevant and insightful responses. Additionally, don't shy away from iterative refinement. If the initial response doesn't quite hit the mark, rephrase your question or add more context. This iterative process improves the quality of the responses and enhances your skill in interacting with AI, making you more adept at extracting useful information.

There are also common misconceptions about AI capabilities that are worth addressing. One such myth is that AI, like ChatGPT, can make decisions for us. While it's true that AI can analyze data and offer recommendations, the final decision-making rests with us. AI lacks the intuition and emotional intelligence that humans bring to the table, which is crucial in many decision-making scenarios. Another misconception is equating AI intelligence with human intelligence. While ChatGPT can process and generate language in a sophisticated manner, it doesn't possess understanding or consciousness. It's a tool that processes input and generates output based on patterns it has learned from vast amounts of data. Keeping these distinctions in mind helps set realistic expectations and ensures that we use AI as a supportive resource rather than a substitute for human insight.

Recognizing both the strengths and limitations of ChatGPT allows you to use it more strategically. It becomes a companion in creativity and problem-solving, providing assistance where it excels while knowing when to rely on human expertise for areas that require a nuanced understanding. This balanced approach enhances the effectiveness of your interactions with AI and empowers you to navigate the ever-evolving landscape of technology confidently.

1.6 Recognizing ChatGPT's Strengths and Weaknesses

ChatGPT is like a versatile tool in your digital toolbox, ready to tackle a wide range of conversational topics. Whether you're exploring the depths of philosophical debates or simply inquiring about the weather, ChatGPT adapts smoothly. Its strength lies in this flexibility, which makes it possible to engage in discussions that span various subjects with ease. This versatility makes it a valuable asset in both professional and personal settings. For someone juggling multiple roles, from project manager to parent, having an AI capable of switching gears as quickly as you do is incredibly helpful. It can assist with drafting emails in one moment and offer book recommendations in the next.

Another impressive feature is ChatGPT's ability to learn and adapt over time. It refines its responses based on interactions, becoming more attuned to your preferences and style. This isn't just about getting smarter; it's about becoming more in sync with how you think and communicate. This adaptability is akin to having a colleague who's constantly learning and improving, ready to offer better solutions and insights as time goes on. Over time, AI has become more than just a tool; it's almost like a partner in conversation, one that understands your needs and anticipates your questions.

However, like any tool, ChatGPT has its limitations. Understanding sarcasm, for instance, poses a challenge. Sarcasm relies heavily on tone and context, elements that are difficult for AI to grasp fully. Without these cues, the AI might miss the mark, interpreting sarcasm as a straightforward statement. This can lead to responses that feel out of tune with the intended humor or irony. Similarly, maintaining long-term context in conversations is another area where ChatGPT can struggle. While it performs well in short exchanges, keeping track of details across extended dialogues can be complex. This limitation means that sometimes, it might not recall earlier parts of a conversation, requiring you to provide reminders or additional context.

To make the most of ChatGPT's strengths, it's wise to use it for tasks where it excels. For rapid information retrieval, the AI is unmatched. You can ask it for quick facts or summaries, and it will deliver with speed and accuracy. This can save you time, especially when juggling several tasks and needing concise answers quickly. Structuring complex queries for clarity is another way to leverage ChatGPT effectively. By breaking down your questions into clear, straightforward parts, you help the AI provide more accurate and relevant responses. This approach is instrumental in fields like research, where precision is key.

For its weaknesses, human oversight can make a significant difference. By reviewing and refining AI-generated responses, you can ensure the accuracy and relevance of the information. This is especially crucial in areas where nuance and detail matter most. Iterative questioning also plays a vital role in compensating for areas where the AI might fall short. If the initial response isn't quite right, don't hesitate to rephrase or expand upon your question. This iterative process allows you to guide the AI toward more precise answers, enhancing the quality of your interactions.

In the end, recognizing where ChatGPT shines and where it needs a little help turns it into a truly powerful tool. It’s about using its strengths to your advantage while being mindful of its limitations. With a thoughtful approach, ChatGPT can become an invaluable resource in your everyday life, supporting a wide range of tasks and making your interactions more productive and enjoyable. As you explore what ChatGPT can do, remember that it's not just about getting answers—it's about engaging in conversations that enhance your understanding and broaden your perspectives.

Chapter Two

Crafting Effective Questions

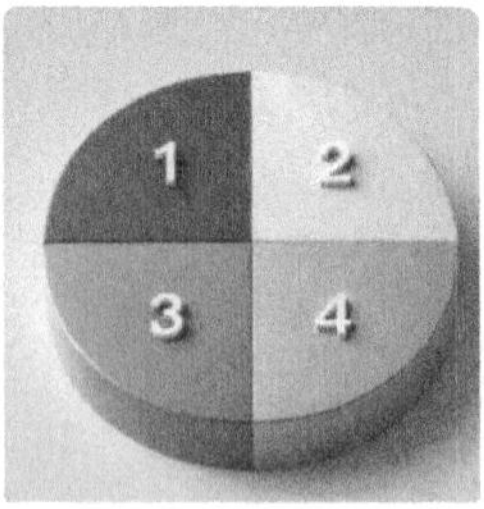

Let me take you back to a moment when I was faced with a challenge that seemed insurmountable. I was preparing a presentation and needed a fresh perspective to make my points resonate. After several failed attempts at brainstorming, I decided to turn to ChatGPT. I typed in a question, hoping for a spark of inspiration. Surprisingly, the response was insightful and offered a new angle I hadn't considered. This experience taught me the power of asking the right questions. It wasn't just about getting any answer; it was about getting a response that propelled me forward. It's this art of crafting effective prompts that we'll explore in this chapter, unlocking the potential for more thoughtful AI interactions.

Prompt engineering is the art and science of crafting inputs that coax the best possible responses from AI models like ChatGPT. Think of it as the difference between asking a friend, "How was your day?" versus "What was the highlight of your day?"

The latter draws out a more meaningful conversation. In the realm of AI, clear and concise prompts are crucial. They guide the AI to focus on what's important, filtering out noise and honing in on the specifics that matter. The phrasing of your question can make all the difference in the quality of the AI's response. Prompt engineering involves understanding and using this dynamic to our advantage, ensuring the AI interprets our queries accurately and provides valuable insights.

Successful prompt engineering requires a mix of specificity and strategic phrasing. Specific language acts as a roadmap, directing the AI to the precise information or insight you seek. For example, instead of asking, "How can I improve my business?" you might say, "What are three strategies to increase customer retention for a small online bookstore?" This specificity clarifies your intent and narrows the scope, making the AI's task more manageable. Incorporating keywords related to your topic further refines the AI's focus, much like adding spices to a dish to enhance its flavor. These keywords act as signals, telling the AI which aspects of the topic are most relevant to your query.

The anatomy of an effective prompt is built on clarity and focus. These components ensure the AI knows exactly what you're asking and why it matters. Clarity is about stripping away ambiguity making sure your question is straightforward and easy to understand. Focus involves aligning your prompt with the desired outcome, like aiming a camera lens at a specific subject to capture the perfect shot. Together, these elements form the backbone of a well-structured prompt, enabling the AI to deliver relevant and insightful responses.

Practice is key to getting better at prompt engineering. Think of it as learning a musical instrument. The more you play, the more intuitive it becomes. Start by experimenting with different scenarios, varying your prompts to see how the AI responds. This exploration helps you understand what works and what doesn't, allowing you to refine your technique over time. Learn from example prompts, analyzing their structure and content to identify successful patterns. As you practice, you'll develop an instinct for crafting prompts that unlock the full potential of AI, turning every interaction into an opportunity for discovery and growth.

2.1 Interactive Element: Prompt Crafting Challenge

Try crafting prompts for these scenarios:

1. **Business Strategy Session:** Ask ChatGPT for innovative marketing ideas for a local coffee shop.
2. **Creative Writing:** Request a plot outline for a mystery novel set in a futuristic city.
3. **Personal Development:** Inquire about effective time management techniques for balancing work and family life.

Note the AI's response for each scenario and adjust your prompt for clarity and specificity. Reflect on what changes improved the quality of the response.

2.2 Avoiding Common Question Pitfalls

It's all too easy to trip up when crafting questions for ChatGPT, especially if you're new to the game. One of the most frequent mistakes is asking overly broad questions. Imagine going into a library and asking the librarian, "Tell me about history." The librarian would be stumped! Similarly, when you toss a vague question to ChatGPT, the AI has to sift through a vast ocean of possibilities. The result? A generic, superficial response that lacks depth and relevance. Another common pitfall is ignoring context. Without context,

your question might be like talking to someone at a party without introducing yourself first. The AI might misinterpret your intent, leading to answers that miss the mark entirely. It's like trying to solve a puzzle without all the pieces—frustrating and ineffective.

The impact on AI responses can be significant when you make these mistakes. Poorly formulated questions often lead to generic and vague answers. ChatGPT may provide an equally aimless answer if your question lacks direction, leaving you unsatisfied. It's not that the AI is incapable; it's more about how the question guides—or doesn't guide—the AI's thought process. Misinterpretation of intent is another consequence of ignoring context. ChatGPT might take your question in a completely unintended direction without the right cues. This can be particularly problematic if you're seeking specific advice or solutions. The AI might latch onto the wrong aspect of your query, providing irrelevant details instead of the insightful guidance you were hoping for.

So, how do you overcome these pitfalls? Focus on clarity and precision. Think of each question as a spotlight, shining brightly on the exact information you seek. Be specific about what you want. Instead of asking, "How do I improve my skills?" try, "What are three online courses for improving coding skills in Python?" This approach narrows the scope, making it easier for ChatGPT to hone in on the most relevant information. Another technique is using qualifiers to narrow down responses. Adding qualifiers like "for beginners" or "in a corporate setting" provides additional layers of specificity, guiding the AI to tailor its response to your unique needs. This enhances the answer's relevance and ensures the information you receive is actionable and valuable.

Let's look at some real-world examples to illustrate these points. Consider the question, "How do I get better at writing?" It's broad and leaves much to interpretation. By refining it to "What are effective writing exercises for improving creative storytelling?" you transform the query into something much more specific. The AI can now focus on providing targeted exercises directly applicable to your goal, such as free writing or character development activities. Another example might involve a work scenario. Asking "How can I be more productive?" could yield a generic list of productivity tips. But by specifying, "What productivity tools can help manage remote team projects?" you're guiding ChatGPT to offer solutions that are directly relevant to your work environment, like suggesting task management or communication apps.

Recognizing and avoiding these common pitfalls enhances your interactions with ChatGPT, making them more effective and rewarding. It's about steering the conversation in a direction that brings out the best in the AI, ensuring that every question you ask is a step towards gaining the insights you need. Remember, the goal is to transform your queries from broad and aimless to precise and purposeful, turning each interaction into a productive dialogue that delivers the answers you seek.

2.3 Using Context for Tailored Responses

Think about a conversation with a friend where they know just the right thing to say because they understand the background of your situation. Context plays a similar role in interactions with ChatGPT. Providing context is like setting the stage for a play; it gives the AI the backdrop it needs to offer relevant and meaningful responses. When you include background details, you enhance the specificity of your query, making it easier for ChatGPT to hone in on the precise information you're seeking. This specificity transforms a simple question into a tailored request that considers the nuances of your situation. For instance, if you're asking for dinner ideas, mentioning dietary preferences or the type of cuisine you're interested in allows the AI to suggest options that are more aligned with your tastes.

Incorporating context effectively involves sharing relevant information upfront. This might mean including details about the situation or the desired outcome in your initial prompt. By doing so, you equip ChatGPT with the necessary information to generate a response that's accurate and aligned with your expectations. Another technique is linking questions to previous interactions. This is particularly useful in ongoing conversations where maintaining context is key. By referencing earlier exchanges, you can ensure continuity and coherence, allowing ChatGPT to build on what has already been discussed. This approach mirrors how we naturally weave past conversations into our current dialogue with others, adding depth and richness to the interaction.

Consider a scenario where you're using ChatGPT for professional advice. You might personalize your query with user-specific details, such as your role, industry, or current challenges. For example, asking, "What strategies can I use to improve team collaboration in a remote work environment as a project manager?" provides a wealth of context that guides the AI's response. In another example, contextualizing professional scenarios by including industry-specific jargon or frameworks helps ChatGPT deliver insights that directly apply to your field. This level of detail enhances the response's relevance and demonstrates the AI's capability to provide advice to your unique circumstances.

However, maintaining context in dynamic conversations can present challenges. In multi-turn dialogues, it's easy for the AI to lose track of the context established earlier. To mitigate this, you can use techniques for context continuity. This might involve summarizing key points from previous interactions at the start of each new query or explicitly referencing past discussions. Doing so, you help ChatGPT stay aligned with the conversation's trajectory, ensuring that its responses remain pertinent and informed. Another challenge is ensuring that the AI doesn't drift off-topic. Managing this requires a balance between providing enough context to guide the conversation and avoiding unnecessary details that might lead to confusion. This balance is akin to steering a discussion with a friend, where you provide enough background to keep them engaged and informed without overwhelming them with irrelevant information.

2.4 Specificity vs. Generality: Finding the Balance

When crafting questions for ChatGPT, finding the sweet spot between specificity and generality is like tuning a radio to get the clearest signal without losing the richness of the sound. Specificity offers clarity, ensuring the AI zeroes in on the precise information you're after. This is especially useful when you need detailed advice or in-depth answers. However, focusing too narrowly might limit the AI's ability to provide a broader perspective, and you might miss out on alternative insights. On the other hand, generality opens the floor to a wide range of responses, which can be beneficial if you're exploring new ideas or seeking inspiration. The challenge is to achieve clarity without boxing in the AI, allowing it to focus on your request while still offering a variety of perspectives. Balancing these elements ensures that your questions lead to responses that are not only accurate but also rich in content and context.

Determining the right level of specificity involves assessing the complexity of the topic at hand. If the subject is straightforward, like asking for the weather forecast, a specific question yields the best results. However, for complex issues, such as exploring business strategies, a touch of generality might invite fresh ideas and cover more ground. Evaluate the desired depth of response you need. If you want a comprehensive answer, lean towards specificity to guide the AI. For exploratory discussions, keeping the question a bit more general could be advantageous, inviting a broader range of insights without overwhelming the AI with constraints. It's about gauging what you want from the interaction and adjusting your question to meet that need best.

Imagine you're planning a vacation and want some ideas. A broad prompt like "Tell me about vacation spots" might fetch general information about popular destinations. But if you refine it to "What are the best family-friendly vacation spots with outdoor activities in the Midwest?" you'll likely get suggestions tailored to those criteria. Specificity directs the AI to hone in on relevant details, while a general approach keeps possibilities open, allowing for a more exploratory conversation. Adjusting specificity based on context plays a crucial role. In a work setting, asking, "What are some productivity tools?" could lead to various software suggestions. However, specifying "What are productivity tools for managing remote teams?" narrows the field to those most applicable to your situation. This adjustment ensures the AI response is relevant and practical, immediately providing useful solutions.

The benefits of balanced questioning are numerous and varied. It facilitates comprehensive and insightful answers by giving the AI enough information to understand your needs while leaving room for creative solutions. This balance helps explore diverse perspectives, which can be incredibly valuable when tackling complex problems or seeking innovative ideas. You encourage informative and thought-provoking responses by allowing the AI to interpret your question within a defined yet flexible framework. This approach transforms your interaction with ChatGPT from a simple Q&A session into a meaningful dialogue, where the AI acts as a partner in discovery rather than just a source of information. Through balanced questioning, you enhance the quality and depth of the insights you gain, making every conversation with ChatGPT a richer, more rewarding experience.

2.5 Iterative Questioning Techniques

Imagine you're sculpting a piece of art. Each chisel and tweak brings you closer to the masterpiece you envision. Iterative questioning with ChatGPT works similarly. This technique involves refining your queries through a series of adjustments, enhancing the clarity and precision of AI interactions. It's about viewing each AI response as feedback, an opportunity to improve upon the previous exchange. With each iteration, you narrow down your focus, gradually honing in on the answers you seek. This continuous improvement process not only refines the quality of the AI's responses but also helps you develop a sharper skill in engaging with these powerful tools. The goal is to turn a broad question into a finely tuned inquiry that elicits the most relevant and insightful answers.

The benefits of iterative questioning are twofold. First, it enhances clarity by refining unclear responses. Think of it like adjusting the settings on a camera until the picture comes into focus. If the initial answer from ChatGPT seems vague or off-point, you can rephrase your question, add more details, or even break it down into smaller parts. This approach gradually sharpens the AI's understanding, providing more accurate and specific answers. Second, iterative questioning allows you to build upon previous answers. Each response from the AI becomes a stepping stone, guiding you toward more complex or nuanced inquiries. This method of layering questions helps create a dynamic dialogue, where each exchange builds upon the last, leading to a more comprehensive understanding of the topic at hand.

Consider a scenario where you're exploring new marketing strategies for your business. You might start with a broad question like, "What are innovative marketing strategies?" The AI could provide a list of general ideas, but to refine this further, you might ask, "How can I apply guerrilla marketing tactics for a local café?" By iterating on your initial question, you tailor the AI's response to suit your specific needs. This step-by-step refinement transforms a complex query into manageable parts, enhancing the relevance and usefulness of the information received. Adjusting questions based on AI feedback is crucial. If a response seems too general, probe deeper by asking follow-up questions or requesting examples. This iterative process not only clarifies the information but also reveals new angles and insights you might not have considered initially.

Incorporating iterative questioning into your interactions with ChatGPT involves setting clear goals for each iteration. Before you start, think about what you hope to achieve with your question. Are you looking for a list of ideas, a detailed explanation, or a creative suggestion? Defining your objective helps guide the refinement process, ensuring that each iteration brings you closer to your desired outcome. Analyzing the AI's responses is equally important. Pay attention to what works and what doesn't. If a particular phrasing yields better results, consider why that might be and apply similar techniques to future questions. This analysis is like a feedback loop, continually improving your ability to craft effective prompts.

Interactive Element: Iterative Questioning Exercise

Try engaging in a dialogue with ChatGPT using iterative questioning. Start with a broad question, such as "What are some ways to improve team collaboration?" Take note of the response and refine your question to add more detail or context. For example, follow up with, "How can remote teams enhance collaboration using digital tools?" Continue this process, adjusting your questions based on the AI's feedback, and observe how the quality of responses improves. Reflect on how each iteration brought you closer to the information you were seeking.

Iterative questioning is a technique and a mindset that encourages curiosity and exploration. It transforms your interaction with AI from a one-time exchange into an ongoing conversation, allowing you to delve deeper into topics and uncover valuable insights. By embracing this approach, you enhance your ability to engage with ChatGPT effectively, making each interaction a step toward greater understanding and discovery.

Where are you in the 4 stages of learning and skill acquisition?

2.6 Building a Question-Formulation Framework

Creating a "question-formulation" framework is like building a solid foundation for a house. It provides consistency and reliability, ensuring that every question you ask ChatGPT is well-crafted and purposeful. A structured framework serves as a guide, helping you articulate your queries with precision and intent. It's about establishing a set of principles that you can rely on, much like a roadmap, to navigate the complex terrain of AI interactions. Having a framework in place ensures that your questions consistently hit the mark, eliciting relevant and insightful responses. This consistency improves the outcomes of your interactions and enhances your confidence in using AI as a reliable partner in problem-solving and creativity.

Key elements like **clarity, context**, and **specificity** are at the heart of an effective question-formulation framework. Clarity ensures that your questions are straightforward and easy to understand, minimizing the risk of misinterpretation. Context adds depth to your queries, providing the background information necessary for ChatGPT to tailor its responses to your needs. Specificity narrows the focus, guiding the AI to hone in on the exact details you're interested in. Together, these elements form a cohesive structure that supports effective questioning.

Additionally, **iteration** and **adaptation** are crucial components. They enable you to refine your framework, adapting to new insights and experiences. This adaptability ensures that your framework remains dynamic and responsive, capable of evolving alongside your growing expertise in AI interactions.

Building your question-formulation framework is a step-by-step process that begins with identifying common question types. Consider the kinds of inquiries you frequently make, whether related to work, personal interests, or creative projects. This categorization helps you recognize patterns and tailor your framework to address specific needs. Next, establish criteria for prompt evaluation. This involves defining a successful query, such as its ability to elicit detailed, accurate, and relevant responses. By setting clear benchmarks, you create a standard against which you can measure the effectiveness of your questions, allowing you to continuously refine and improve your framework.

To illustrate the power of a well-structured framework, let's explore its application in professional settings. Imagine a team of marketing professionals using a framework to streamline their brainstorming sessions. By standardizing their approach to questioning, they ensure that their prompts consistently yield high-quality insights, enhancing their creative output. This framework might include specific guidelines for incorporating audience demographics or campaign objectives into their queries, ensuring the AI's responses align with their strategic goals. In another scenario, a software development team might adapt their framework to focus on technical details, guiding ChatGPT to provide code snippets or debugging tips. These examples highlight the versatility of a question-formulation framework, demonstrating its ability to enhance AI interactions across various use cases.

By sharing examples of successful frameworks in use, we see how they can be adapted to different contexts. Whether you're in a corporate environment or pursuing personal projects, a framework provides the structure needed to maximize the benefits of ChatGPT. It's about creating a system that supports your unique needs, enabling you to engage with AI confidently and effectively. As you develop your framework, remember it's not set in stone. Allow it to evolve and adapt as you gain new insights and experiences. This flexibility ensures that your framework remains a valuable tool, capable of guiding you toward more meaningful and productive AI interactions. With a robust framework in place, you're well-equipped to explore the limitless possibilities that ChatGPT offers, transforming your approach to AI from **reactive to proactive** and opening the door to a world of creative and intellectual exploration.

Chapter Three

Overcoming Barriers to Effective AI Use

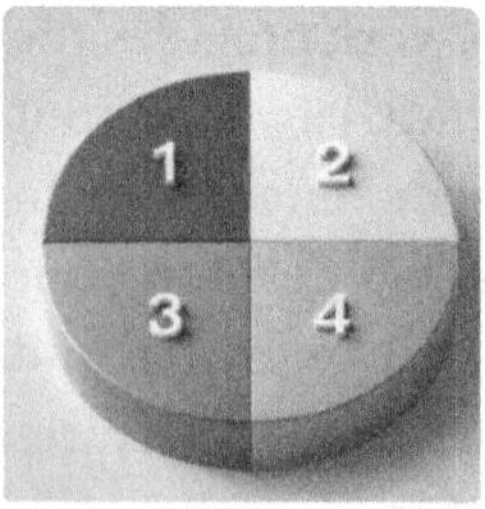

Imagine you're at a dinner party, and the conversation shifts to artificial intelligence. You nod along, but inside, you're scrambling to keep up with the terms flying around the room. Words like "neural networks" and "machine learning" seem as foreign as another language. I remember feeling the same way when I first started exploring AI. It was like opening a book to a random page in a foreign language and trying to understand the plot. This chapter aims to demystify that jargon for you, turning AI's complex language into simple, relatable concepts that anyone can grasp.

Understanding AI begins with recognizing the key terms that often come up in discussions. Let's start with "Natural Language Processing," or NLP. Think of it as the tool that allows AI to understand and generate human language. It's the reason why your virtual assistant can interpret your request to play a song or why a chatbot can answer your queries. It's as if the AI has learned our language, enabling it to communicate effectively. Then there's "Machine Learning," which is essentially how AI learns from data. It's like teaching a child to recognize animals by showing them pictures repeatedly until they can identify a dog from a cat. In AI, this process involves algorithms that sift through data, identifying patterns and improving with each iteration.

To make these concepts more relatable, consider neural networks. These are the backbone of machine learning, mimicking the human brain's neuron connections. Imagine a neural network as a busy highway system where information travels rapidly between different points, helping AI process tasks like speech recognition and image analysis. Algorithms, on the other hand, are like recipes for problem-solving. Just as a recipe guides you in preparing a dish, algorithms guide AI in processing data to reach conclusions. They are the step-by-step instructions machines follow to perform tasks efficiently and accurately.

Jargon exists in AI for a reason. It facilitates precise communication in technical discussions. Like any specialized field, AI relies on specific terms to succinctly convey complex ideas among professionals. This precision is crucial when developing technologies and ensuring they work as intended. However, it can be a barrier for newcomers trying to understand AI's potential. By breaking down this jargon into simpler terms, we can make AI more accessible and less intimidating for everyone.

By breaking down these complex concepts, you'll find that AI isn't as daunting as it seems. It's about turning the unfamiliar into the familiar, making AI a tool you can use confidently and effectively. Whether discussing AI at a party or exploring its potential in your projects, understanding this language opens up new possibilities and empowers you to engage with technology more meaningfully.

3.1 Building Confidence: Overcoming Fear of Mistakes

Imagine approaching ChatGPT like you would a new dance class—excited but hesitant, unsure if your feet will follow the rhythm. That initial fear of making mistakes can be a significant barrier to effectively using AI. It's natural to feel apprehensive when faced with something new and complex. This fear can lead to avoidance, where you might hesitate to explore all the possibilities ChatGPT offers, sticking to what feels safe and familiar. You might be reluctant to experiment, worried about receiving incorrect or nonsensical responses. This reluctance can result in missed opportunities to discover AI's full potential, like a dancer who never leaves the comfort of simple steps.

So, how do you overcome this fear and build confidence with AI? It starts with taking small, manageable steps. Begin by engaging ChatGPT with simple questions or tasks. Think of these interactions as warm-up exercises, gradually building your comfort level. Small successes will boost your confidence, making you more willing to tackle more complex queries. It's also crucial to embrace the idea that mistakes are a natural part of the learning process. Each misstep is not a failure but a stepping stone toward improvement. When ChatGPT gives you an unexpected answer, view it as an opportunity to refine your question or explore a different angle. This perspective shift transforms errors into valuable lessons, reinforcing your understanding and skill in using AI.

Consider the story of a colleague who was once intimidated by ChatGPT. Initially, she struggled with crafting effective prompts, often receiving irrelevant answers. But instead of giving up, she treated each interaction as a learning experience. Slowly, through trial and error, she began to see patterns in how the AI responded to different types of questions. Over time, her confidence grew, and she started using ChatGPT to brainstorm innovative ideas for her projects. What once seemed daunting became integral to her workflow, highlighting the transformative power of persistence and an open mind. This journey from apprehension to mastery is a testament to what's possible when we confront our fears and view challenges as opportunities for growth.

Adopting a growth mindset is essential in this process. It's about celebrating progress over perfection and understanding that every interaction with AI is a chance to learn something new. When you view errors as opportunities for improvement rather than setbacks, you create a positive feedback loop that encourages continuous learning. This mindset applies to AI and enriches other areas of life, fostering resilience and adaptability. Embracing this approach helps you overcome the fear of making mistakes, transforming it into a catalyst for development and creativity. By recognizing that perfection isn't the goal, you free yourself to explore, experiment, and ultimately, grow.

You might also take inspiration from those who have found success after overcoming initial hurdles with AI. One user shared how ChatGPT became a critical tool in her daily routine after an initial period of trial and error. At first, she was overwhelmed by the technology, unsure how to make it work for her unique needs. However, she persisted, gradually learning to tailor her questions and effectively harness the AI's capabilities. Today, she marvels at how far she's come, crediting her willingness to learn from mistakes as the key to her success. Her story illustrates the power of perseverance and the rewards of embracing a challenge head-on.

When you approach ChatGPT with a mindset geared toward growth, each interaction becomes a stepping stone toward greater understanding and proficiency. Encouraging small, manageable interactions with the AI while viewing mistakes as learning opportunities can significantly boost your confidence. Embracing the learning potential of errors allows you to develop a more nuanced understanding of how to engage with AI effectively. By cultivating a growth mindset, you transform the experience of using AI from a source of anxiety to an empowering tool for creativity and problem-solving.

3.2 Handling Inconsistent Responses: A Guide

Have you ever asked ChatGPT the same question twice and received different answers each time? You're not alone. This variability in responses can be puzzling, especially when you expect consistency. The root of this inconsistency often lies in the language model's inherent design. ChatGPT, like other AI models, doesn't always provide the same answer because its outputs are influenced by the vast array of data it has processed. This variability is akin to asking two people the same question—they might draw from different experiences or interpretations,

leading to varied responses. Additionally, ambiguous or unclear prompts can further compound this inconsistency. When a question lacks specificity or context, the AI might interpret it in multiple ways, each leading to a different answer. This is why clear and concise communication is crucial.

To manage these inconsistencies, refining your prompts is key. Think of it as the difference between a vague request and a detailed order at a restaurant. The more specific you are, the more likely you will get the desired result. Start by identifying what you need from the AI. Are you looking for a step-by-step guide or a simple yes-or-no answer? Structuring your question with this end in mind can help guide the AI's response. Using context to guide your questions further refines the AI's focus. Including relevant background information helps ChatGPT understand the scope of your inquiry, reducing the chances of receiving a response that veers off course. This technique is like setting a GPS destination before starting a journey—it ensures you're heading in the right direction.

Let's explore some scenarios where inconsistencies might arise. You might ask, "What's the best way to handle stress?" and receive a response about meditation the first time, then suggestions for exercise the next. Both are valid, but lacking a specific focus can lead to varied advice. Another example might involve asking for marketing strategies. One day, ChatGPT might suggest social media campaigns, while another day, it might focus on email marketing. These differences highlight how the AI interprets your question based on subtle nuances in phrasing or context. Analyzing these differences can offer insights into how to phrase your questions more effectively. It's like fine-tuning a radio dial to get the clearest signal—minor adjustments can make a big difference.

Refining your approach involves iterative questioning. This technique allows you to build on previous interactions, gradually honing in on the most relevant information. Start with a broad question, then use follow-up queries to narrow the focus. Each iteration provides an opportunity to clarify and specify, guiding the AI to a more precise answer. Seeking feedback is another valuable strategy. Whether from peers or through analyzing the AI's responses, feedback helps you identify areas for improvement. Consider discussing your experiences with others who use ChatGPT. Their insights can highlight new ways to phrase questions or approach problems. This collaborative approach improves your questioning techniques and effectively enriches your understanding of leveraging AI.

Focusing on clarity and context allows you to transform your interactions with ChatGPT from hit-or-miss exchanges into productive dialogues. This process enhances the quality of the AI's responses and empowers you to engage more confidently and effectively with technology.

3.3 Navigating AI's Limitations with Confidence

Understanding the limitations of ChatGPT is akin to knowing the quirks of a well-loved car. You trust it, but you also know when it might need a little extra care on the road. One of the main challenges AI faces is grasping the subtleties of nuanced language. Imagine trying to explain a joke to someone who doesn't get sarcasm; sometimes, that's how ChatGPT interacts with our more layered expressions. While it excels at processing straightforward requests, it can stumble over language that relies on tone, context, or cultural references to convey meaning. Another hurdle is maintaining long-term context within conversations. ChatGPT, despite its vast data-driven insights, can lose track of the thread if the dialogue becomes too extended or complex. It's like trying to remember a conversation from a week ago without any notes—some details might slip through the cracks.

But fear not because there are practical ways to work around these limitations. When dealing with complex inquiries, think of ChatGPT as one tool in your toolbox. Supplement it with other resources, like research papers, expert opinions, or specialized databases, to fill in the gaps. This approach is similar to consulting multiple references when tackling a challenging project. Setting realistic expectations for what AI can handle is equally essential. It's about knowing when to leverage ChatGPT's strengths and when to seek human insight. Recognizing the boundaries of AI capabilities helps you use it more effectively, ensuring you aren't left frustrated when the technology doesn't meet every need. By clearly defining ChatGPT's role in your tasks, you can create a balanced workflow that maximizes the strengths of AI and human expertise.

The collaboration between AI and humans is where the magic truly happens. Think of AI as your diligent assistant, tirelessly processing data and offering insights while you, the human, bring creativity and intuition to the table. AI can augment decision-making by providing quick data analysis or generating options you might not have considered. Meanwhile, your human insight can refine these suggestions, adding a layer of judgment that AI cannot replicate. It's a harmonious partnership where both parties enhance each other's capabilities, much like a chef and sous chef working together to create a culinary masterpiece. This collaboration underscores the importance of not viewing AI as a replacement but as a complementary force that enhances what humans do best.

Proactive learning and adaptation are key to navigating AI effectively. Staying informed about AI advancements allows you to understand and anticipate technological changes. Consider it like keeping up with the latest trends in your industry—knowledge that keeps you competitive and prepared. Adapting your techniques based on AI feedback is another crucial element. As you interact with ChatGPT, note what works and what doesn't. Use this information to refine your approach, much like a gardener who adjusts their methods based on how their plants respond. This continuous improvement mindset ensures you're constantly evolving and ready to leverage new AI capabilities as they become available.

Textual Element: Adaptive Learning Checklist

- Identify AI's strengths and use them for routine tasks.
- Supplement AI insights with human expertise for complex problems.
- Stay updated on AI developments through reliable sources.
- Adjust your questioning techniques based on AI feedback and results.

Use this checklist as a guide to optimize your AI interactions, ensuring you're always learning and adapting to get the best results.

You can confidently navigate the AI landscape by acknowledging ChatGPT's limitations and embracing its strengths. It's about understanding the technology's place within your toolkit and leveraging it to enhance your decision-making and creativity. With a thoughtful approach, you can turn what might seem like obstacles into opportunities for growth and innovation, creating a partnership between humans and machines that is as powerful as it is effective.

3.4 Addressing Skepticism: How AI Can Meet Your Needs

Skepticism around AI is more common than you might think, and it's only natural, given the rapid pace of technological change. One major source of doubt stems from concerns about privacy and data security. Many people worry about how their personal information is used and whether AI systems are adequately protected from breaches. These concerns are valid, especially in a world where data is often referred to as the new oil. Misunderstandings about AI capabilities further fuel skepticism. Some folks might see AI as a magical solution to all problems, while others view it as a potential threat to jobs and privacy. Both extremes miss the mark, as AI is neither an all-powerful savior nor a menace lurking in the shadows. It's a tool, one that, when used correctly, can enhance our lives in numerous ways.

To understand AI's practical benefits, consider its role in automating repetitive tasks. Imagine the hours saved when AI handles mundane tasks like sorting emails or scheduling meetings. This automation frees up time for more creative and strategic work, enhancing productivity across the board. There are countless success stories where AI-driven solutions have transformed industries. Take, for example, a logistics company that uses AI to optimize delivery routes, reducing fuel costs and improving efficiency. Or consider a healthcare provider using AI to analyze patient data, leading to faster diagnoses and better treatment plans. These examples highlight AI's potential to solve real-world problems and improve outcomes in various fields.

Transparency is key to addressing skepticism. We can alleviate privacy concerns by clearly explaining AI's data handling practices. For instance, many AI systems employ robust encryption and strict access controls to protect data integrity. It's also essential to clarify AI's limitations and potential. While AI excels at processing large datasets and identifying patterns, it lacks human intuition and emotional intelligence. Understanding these boundaries helps manage expectations and ensures that AI is seen as a complement to human skills, not a replacement. Open and honest communication about what AI can and cannot do builds trust and fosters a more informed perspective.

Encouraging an open-minded approach to AI involves promoting a willingness to experiment and learn. Start by using AI for simple tasks, like setting reminders or generating quick summaries. As you become more comfortable, you can expand its use to more complex applications, such as analyzing data trends or brainstorming new ideas. This gradual exploration allows you to build confidence in AI's capabilities while gaining firsthand experience with its limitations. It's like learning to ride a bike, beginning with training wheels before tackling a mountain trail. By taking small steps, you can appreciate AI's value to your routine, discovering new ways to integrate it into your personal and professional life.

Textual Element: Quick-Start AI Challenge

Try incorporating AI into a daily task. Choose something simple, like using a voice assistant to set reminders or asking ChatGPT for meal ideas. Reflect on the experience: What worked well, and what challenges did you encounter? Consider how you might expand AI's role in your routine. This exercise helps build familiarity and confidence, turning AI from an abstract concept into a practical tool you can rely on.

You can unlock AI's full potential by addressing skepticism with transparency and encouraging exploration. It's about understanding the nuances of this powerful tool and recognizing how it can meet your needs. With an open mind and a willingness to learn, AI can become an invaluable ally in your everyday tasks and larger projects, enhancing efficiency, creativity, and problem-solving capabilities.

3.5 Time-Saving Strategies for Busy Professionals

In today's fast-paced world, finding ways to save time is more critical than ever. With the help of AI tools like ChatGPT, you can significantly streamline your daily workflow. Picture this: you start your day with a barrage of emails that need sorting. Instead of spending precious time manually sifting through them, you let AI handle it. ChatGPT can help automate routine communications by categorizing emails based on urgency or subject matter, allowing you to focus on what truly matters. This simple yet effective use of AI can free up hours in your week, letting you dedicate more energy to strategic tasks.

One of the most impressive aspects of AI is its ability to enhance information retrieval speed. Need to pull together a quick report or find the latest data on a topic? ChatGPT can do that in seconds. It's like having an assistant who never sleeps, always ready to give you the information you need at lightning speed. Imagine preparing for a meeting and needing a quick overview of current trends in your industry. With AI, you can access concise summaries that keep you informed without requiring lengthy research sessions. This capability boosts your productivity and always prepares you with the latest insights.

Integrating AI into your daily tasks doesn't have to be daunting. Start by setting up automated scheduling reminders. This is a simple yet effective way to ensure you never miss an appointment or deadline. Let AI handle these routine reminders, freeing your mind to concentrate on more pressing matters. Also, consider using AI for quick research and summaries. Whether condensing a long document into key points or generating a brief overview of a complex subject, AI can provide valuable assistance. These small integrations can transform how you manage your day, making your workflow more efficient and less stressful.

Let's explore some real-world examples of AI efficiency in action. Imagine a busy executive who uses AI-powered email sorting to prioritize messages that require immediate attention. By delegating this task to ChatGPT, they can focus on high-level decision-making without getting bogged down by an overflowing inbox. Similarly, automated report generation is another area where AI shines. Instead of manually compiling data and drafting reports, professionals can rely on AI to gather and organize information, producing polished documents ready for presentation. These examples illustrate how AI can be a game-changer in managing daily tasks, allowing you to operate efficiently.

However, it's crucial to strike a balance between AI and human tasks. While AI excels at handling repetitive and time-consuming activities, human input is unparalleled when it comes to strategic planning and creative problem-solving. Delegate tasks like sorting emails and scheduling to AI, but keep the critical thinking and decision-making responsibilities for yourself. This division of labor ensures that you're leveraging AI where it makes the most impact while maintaining the human touch in areas requiring judgment and intuition. It's about using AI to support and enhance your capabilities, not replace them. This balanced approach allows you to maximize productivity while still engaging in the tasks that require your expertise and creativity.

This chapter explored how AI can transform your workday, from automating routine tasks to enhancing your ability to retrieve information swiftly. By incorporating AI into your daily operations, you can save time and focus on the strategic elements of your role. As we move forward, we'll dive deeper into how AI can be personalized to suit individual needs, ensuring that the technology works for you and not vice versa.

Unlock the Power of Generosity

"A single act of kindness throws out roots in all directions." - Amelia Earhart

When we help others, we spark joy—not just for them but for ourselves too. So, why not spread a little happiness today?

Would you help someone like you—curious about ChatGPT but unsure where to start?

My mission with *Unlocking ChatGPT for Beginners* is to make learning this amazing tool simple, fun, and approachable for everyone.

But I need your help to reach more people.

Did you know most people pick books based on reviews? Your opinion matters more than you think! By leaving a review, you could help someone else...

- Take their first step into the exciting world of AI.
- Feel confident using ChatGPT to solve everyday challenges.
- Unlock their creativity and productivity in new ways.
- Make technology work *for* them, not against them.

Your review costs nothing, takes just a moment, and can make a big impact.

How to Leave a Review

Scan the QR code below or click this link:

https://www.amazon.com/review/review-your-purchases/?asin=1967215014

Share a few words about how this book helped you. Your kind words could inspire someone else to start their journey with ChatGPT.

If you enjoy helping others, you're my kind of person. Thank you from the bottom of my heart for taking the time to make a difference!

Thanks,

C.P. Bailey

Chapter Four

Real-World Applications of ChatGPT

"The best way to predict the future is to create it." – Peter Drucker

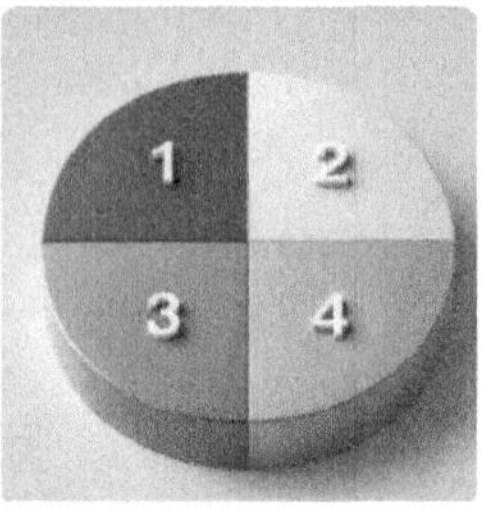

Picture this: it's a Monday morning, and your inbox is brimming with emails that need your attention. You feel the weight of the workweek pressing down already. But what if there was a way to lighten that load, allowing you to focus on more strategic tasks rather than getting bogged down by routine correspondence? Enter ChatGPT—a tool that can revolutionize how you handle repetitive tasks, freeing up your time for what truly matters. Imagine having an assistant that manages emails, schedules appointments, and sends reminders, all with minimal input from you. This is the power of automation that ChatGPT brings to the table, transforming how you work by handling mundane tasks that consume valuable time.

ChatGPT excels in automating routine tasks, such as generating email responses. It can draft replies to common queries, ensuring consistency and accuracy in communication. This capability is particularly valuable for businesses that receive high volumes of similar emails, allowing them to maintain a high standard of customer service without the need for constant human intervention. With AI managing these tasks, your inbox becomes less of a burden, allowing you to concentrate on more critical projects. Similarly, ChatGPT can handle scheduling appointments and sending reminders. By integrating with your calendar, it can automatically book meetings, send confirmations, and remind you of upcoming events. This seamless coordination eliminates the back-and-forth often involved in scheduling, streamlining your workflow, and ensuring you never miss a beat.

Another area where ChatGPT shines is in task prioritization. Creating to-do lists with priority levels helps you organize your day more effectively. You can input tasks and assign urgency levels, allowing ChatGPT to suggest which tasks to tackle first. This feature is akin to having a personal assistant who understands your workload and helps you focus on what's most pressing. By setting deadlines and reminders, ChatGPT also ensures you stay on track, reducing the likelihood of last-minute rushes and overlooked tasks. This proactive approach to time management can significantly enhance productivity, allowing you to allocate your energy and resources efficiently.

ChatGPT's ability to retrieve information quickly is another productivity booster. Whether you need to conduct a quick fact-check or summarize lengthy documents, ChatGPT can provide the necessary information in a fraction of the time it would take manually. Imagine preparing for a meeting and needing a succinct overview of a complex report. ChatGPT can distill the key points, ensuring you're well-informed without having to wade through pages of data. This rapid access to information saves time and enhances decision-making, as you have the insights you need at your fingertips.

AI-driven decision support systems are where ChatGPT truly comes into its own. ChatGPT can offer insights that aid in strategic decision-making by analyzing data trends. It can process large datasets to identify patterns and potential opportunities, providing you with valuable information to guide your choices. This capability is particularly useful in business environments, where data-driven decisions can lead to competitive advantages. Moreover, ChatGPT can generate potential solutions for business challenges by acting as a brainstorming partner that offers fresh perspectives. Presenting various options and scenarios helps you explore different avenues and make informed decisions. This collaborative approach enhances problem-solving, turning ChatGPT into a valuable ally in navigating the complexities of modern business.

Interactive Element: Productivity Boost Checklist

- **Automate Routine Tasks:** Use ChatGPT to handle emails and schedule appointments.
- **Organize Your Day:** Create prioritized to-do lists with deadlines.
- **Streamline Information Retrieval:** Use AI for quick fact-checking and document summaries.
- **Enhance Decision-Making:** Leverage ChatGPT's data analysis to explore solutions.

By following this checklist, you can unlock ChatGPT's full potential, transform your approach to work, and maximize your productivity.

4.1 ChatGPT in Customer Service: Best Practices

Imagine you're a customer trying to resolve an issue late at night, and the company's support line is closed. Frustrating, right? This is where ChatGPT steps in, providing 24/7 customer support. Unlike traditional customer service, which often operates within business hours, ChatGPT is always available. It's like having a diligent team that never clocks out. This continuous availability ensures that queries are addressed promptly, improving customer satisfaction and fostering trust. Whether answering common questions or helping navigate through services, ChatGPT handles these tasks efficiently and accurately. ChatGPT shines by managing multiple interactions simultaneously when dealing with high volumes of inquiries. This capability reduces wait times and alleviates the burden on human agents, allowing them to focus on more complex issues. In turn, businesses can maintain a high standard of service without stretching their resources thin.

Integrating ChatGPT into customer service isn't just about flipping a switch. It requires thoughtful implementation to be truly effective. Training the AI with common customer queries is an essential first step. You ensure the AI can respond accurately and consistently by feeding it a database of frequently asked questions and their corresponding answers. This training is like teaching employees the ropes, preparing them to handle routine inquiries easily.

However, there will be times when a human touch is necessary. Seamless handoff protocols should be in place to transition customers to human agents when needed, ensuring that complex or sensitive issues are handled appropriately. This blend of AI efficiency and human empathy creates a customer service experience that's both responsive and considerate.

Personalization in customer service is another area in which ChatGPT excels. Customers appreciate it when their interactions feel tailored to their needs, and ChatGPT can deliver this by using customer data to inform its responses. Imagine a customer reaching out about their recent purchase; ChatGPT can access their order history and provide updates specific to their situation. This level of personalization makes interactions more meaningful, turning routine exchanges into opportunities for building customer loyalty. Additionally, ChatGPT can implement personalized recommendations based on past interactions, offering suggestions that align with the customer's preferences. This proactive approach enhances the customer experience and increases the likelihood of repeat business.

Case studies provide a window into how businesses successfully leverage AI in customer service. For instance, many e-commerce platforms use ChatGPT to boost sales by offering real-time assistance and recommendations. These platforms enhance satisfaction and drive conversions by guiding customers through their purchasing journey. In the service industry, AI has been instrumental in elevating client satisfaction. For example, a travel company might use ChatGPT to provide instant itinerary updates, ensuring travelers can access the latest information at their fingertips. These examples highlight the transformative impact of AI when integrated thoughtfully into customer service strategies. They demonstrate that with the right approach, ChatGPT can be a powerful tool for enhancing customer interactions and driving business success.

4.2 Using ChatGPT for Creative Brainstorming

Imagine sitting down to brainstorm ideas for a new marketing campaign. The usual methods might involve whiteboards, sticky notes, or endless coffee-fueled meetings. But what if there was a tool that could spark creativity with just a few keystrokes? ChatGPT can do just that, acting as a catalyst for innovative thinking. Generating a wide array of ideas helps you break free from conventional thought patterns. Whether crafting a catchy slogan or developing a concept for a social media blitz, ChatGPT offers fresh perspectives that might not surface in traditional brainstorming sessions. Its ability to provide diverse suggestions based on your input allows you to explore new angles and approaches, making it an invaluable partner in creative processes.

Guiding AI in creative endeavors requires a bit of finesse. It's about knowing how to frame questions that encourage diverse ideas. Start by posing open-ended questions like "What are some unique ways to engage our audience?" This invites ChatGPT to explore various possibilities without being constrained by a narrow focus. Encouraging out-of-the-box thinking is another powerful technique. You might use prompts that challenge conventional norms or introduce unexpected elements, like asking, "How would a child describe our product?" Such unconventional prompts can lead to surprising and innovative insights, pushing the boundaries of what's typically considered.

In collaborative brainstorming, ChatGPT becomes a dynamic team player. It can facilitate virtual sessions by generating ideas that participants can build upon. By having the AI present initial concepts, team members can focus on refining and expanding these ideas, leading to richer outcomes. Human creativity combined with AI-generated suggestions creates a synergy that enhances the overall brainstorming process. For instance, during a remote team meeting, ChatGPT might propose a series of themes or motifs, sparking a lively discussion that results in a cohesive and compelling campaign plan. This collaborative approach speeds up the ideation phase and ensures a wide range of ideas are considered.

Art and design projects provide a canvas where AI's creative potential truly shines. Consider a graphic design team working on a new branding project. Using ChatGPT to generate initial concepts or themes, designers can explore styles and elements they might not have considered. This AI assistance can lead to innovative designs that stand out in a crowded market. Writing workshops also benefit from AI integration. Imagine a group of writers developing a plot for a novel. ChatGPT can suggest plot twists or character developments, offering a fresh perspective that enriches the narrative. By integrating AI into these creative endeavors, teams can push the boundaries of their imagination, producing unique and captivating work.

Interactive Element: Creative Prompt Exercise

- **Challenge:** Use ChatGPT to brainstorm ideas for a new product launch. Start with an open-ended question like "What innovative features could make our product stand out?"

- **Next Steps:** Analyze the AI-generated ideas and select a few to explore further. Combine these with team input to develop a comprehensive launch plan.

This exercise encourages you to experiment with AI in creative settings, discovering its potential to enhance your brainstorming sessions and lead to breakthrough ideas.

4.3 Educational Applications: Enhancing Learning

Imagine walking into a classroom where AI lends a helping hand to the teacher, taking care of the mundane tasks so that more time and energy can be devoted to fostering student creativity and understanding. ChatGPT can be an invaluable teaching assistant, automating tasks like grading and providing assignment feedback. This automation doesn't just save time; it ensures consistency in evaluation, allowing educators to focus on crafting engaging lesson plans and resources. ChatGPT enables teachers to dedicate more attention to individual student needs by handling repetitive assessment tasks and tailoring their approach to fit different learning styles and abilities. This personalized focus enhances the educational experience, making learning more interactive and enjoyable for students.

In the digital classroom, student engagement is key. ChatGPT can elevate this engagement by providing instant answers to student queries, acting as a round-the-clock tutor who never tires. Imagine a student stuck on a homework problem late at night. With ChatGPT, they can get immediate help, breaking down barriers to understanding and keeping the momentum of learning alive.

Additionally, ChatGPT can offer personalized study suggestions based on a student's performance, guiding them toward resources that target their specific areas of improvement. This tailored approach boosts confidence and encourages students to take charge of their learning journey, exploring new topics with curiosity and enthusiasm.

In the realm of remote learning, ChatGPT plays a crucial role in bridging educational gaps. The shift to online education has been challenging for many, but AI can support this transition by enhancing the delivery of courses. It can facilitate interactive tutorials and quizzes, turning passive learning into an active experience. Students can engage with the material at their own pace, revisiting complex topics as needed. This flexibility ensures that no student is left behind, creating an inclusive environment where everyone has the opportunity to succeed. ChatGPT's ability to adapt to different learning environments makes it a versatile tool that supports traditional and modern educational models.

Let's explore some examples of AI-augmented educational programs that have successfully integrated ChatGPT into their curriculum. Language learning platforms, for instance, have harnessed the power of AI to provide practice opportunities that are both engaging and effective. Students can immerse themselves in a new language through interactive dialogues and real-time feedback, gaining confidence and fluency with every interaction. Similarly, AI simulations have enhanced STEM education, allowing students to experiment with complex concepts in a virtual setting. These simulations provide a hands-on experience that deepens understanding and sparks interest in scientific exploration. By incorporating AI into the learning process, educators can create dynamic and stimulating environments that inspire students to pursue their passions and achieve their potential.

Visual Element: AI in Education Infographic

- **Automated Grading and Feedback:** Streamlines the evaluation process, allowing teachers to focus on instruction.
- **Instant Student Support:** Provides 24/7 assistance,

encouraging independent learning and problem-solving.

- **Remote Learning Enhancement:** Supports online education with interactive resources and personalized guidance.
- **Innovative Educational Programs:** Language and STEM initiatives that integrate AI for deeper engagement.

These elements illustrate how ChatGPT transforms education, paving the way for a future where learning is accessible, personalized, and engaging.

4.4 ChatGPT for Professional Development and Career Growth

When it comes to skill enhancement, ChatGPT is like having a personal tutor available at all times. Imagine diving into a new field, brushing up on existing skills, and having AI guide you with personalized learning recommendations. This isn't just about suggesting random courses; ChatGPT can analyze your current skill set and career goals, offering a tailored learning path that aligns with where you want to go. This personalized approach helps you focus on the areas that matter most and ensures that your learning journey is efficient and effective. Beyond recommendations, ChatGPT can critique professional writing, providing feedback that can refine your communication skills. Whether you're drafting reports or crafting emails, AI can highlight areas for improvement, helping you present your ideas clearly and professionally.

In terms of career coaching, ChatGPT acts as a virtual mentor, offering insights and advice to navigate the professional landscape. It can simulate interview scenarios, preparing you for challenging questions and helping you confidently articulate your responses. This practice builds your interview skills, making you more prepared for real-world interactions. Additionally, ChatGPT can create personalized career pathways, mapping out potential steps to reach your professional goals. It identifies opportunities for advancement and suggests strategies to overcome obstacles, providing a comprehensive plan that supports your career growth. This AI-driven guidance empowers you to make informed decisions about your career, ensuring you're always moving forward.

Networking is another area where ChatGPT can make a significant impact. By identifying relevant networking events and suggesting online forums and communities, AI helps you connect with industry peers and expand your professional circle. This isn't just about meeting new people; it's about building relationships leading to collaborations, mentorship, and career opportunities. ChatGPT can point you toward communities that align with your interests and goals, ensuring you're engaging with the right people. This targeted approach to networking enhances your professional visibility and opens doors to new possibilities. By leveraging AI to navigate the networking landscape, you can make meaningful connections that support your career aspirations.

Success stories abound when it comes to individuals leveraging AI for career advancement. Take, for instance, entrepreneurs who use ChatGPT for business planning. By analyzing market trends and generating strategic insights, AI helps these entrepreneurs make data-driven decisions that drive growth and innovation. This AI-assisted planning supports the creation of robust business strategies that position companies for success. Professionals in various fields are also expanding their skill sets with AI-assisted learning. Whether mastering a new software tool or delving into industry-specific knowledge, ChatGPT provides the resources and support needed to stay competitive. These stories illustrate the transformative power of AI in professional development, highlighting how ChatGPT can be a valuable ally in achieving career success.

4.5 Streamlining Remote Work with AI Assistance

Imagine sipping your morning coffee while ChatGPT takes care of the routine admin tasks that usually clutter your day. This AI marvel can manage everything from booking meetings to organizing documents, giving you more time to focus on the big picture. Automating these repetitive tasks is like having a virtual assistant who's always on call. It handles the nitty-gritty details, freeing your brain for strategic thinking. ChatGPT isn't just about doing more; it's about doing better, allowing you to allocate your mental energy to tasks requiring insight and creativity.

Beyond the mundane, ChatGPT is pivotal in facilitating virtual team collaboration. Picture a team spread across multiple time zones yet working seamlessly as if they were sitting in the same room. ChatGPT can coordinate team tasks and timelines, ensuring everyone is on the same page without the hassle of endless email chains. It drafts clear and concise emails and meeting agendas, providing smooth and effective communication. Managing project timelines and updates becomes a breeze, as ChatGPT can track progress and send reminders, keeping the entire team aligned and focused on shared goals. This AI-driven coordination reduces the friction in distributed teams, enhancing productivity and cohesion.

But what about the human side of remote work, the camaraderie and connection that a physical office fosters? ChatGPT steps in here, too, facilitating virtual team-building activities that bring colleagues together, even if they're miles apart. Imagine a virtual trivia night or a collaborative art project, all organized with the help of AI. These activities break the monotony of remote work, fostering team spirit and boosting morale. Moreover, ChatGPT can offer personalized support and resources tailored to individual team members' needs, creating a supportive and inclusive work environment. It's about making remote work more human, ensuring team members feel connected and valued despite the physical distance.

Numerous organizations have already embraced AI to transform their remote work dynamics. Consider remote-first companies that have optimized their operations using AI tools. By integrating ChatGPT into their workflows, these companies have streamlined processes, reduced overhead costs, and enhanced employee satisfaction. AI-driven solutions have enabled startups to scale quickly, providing the flexibility and agility needed to thrive in competitive markets. These companies use ChatGPT to handle everything from customer support to project management, illustrating the transformative power of AI in modern business. With AI as an ally, companies can navigate the challenges of remote work, turning potential obstacles into opportunities for innovation and growth.

As we wrap up this chapter, it's clear that AI, and ChatGPT in particular, is reshaping the remote work landscape. From automating routine tasks to enhancing team communication and morale, AI offers solutions that make remote work possible and highly effective. This chapter has explored how AI can be a powerful tool in the remote work toolkit, supporting productivity and connectivity in an increasingly digital world. As we move into the next chapter, we'll explore the ethical considerations and responsibilities of integrating AI into our professional lives.

Chapter Five

Interactive Learning and Continuous Improvement

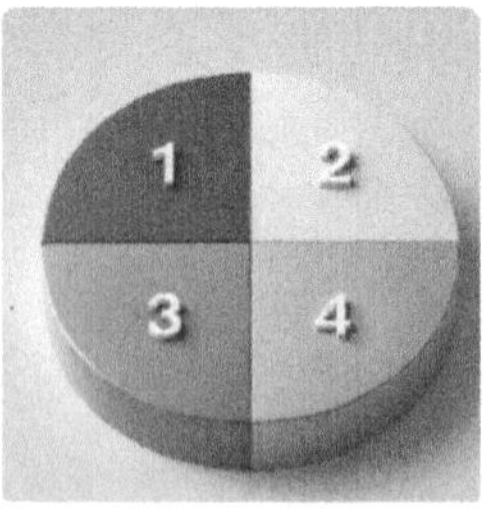

Imagine standing at the edge of a blank canvas, brush in hand. The possibilities are endless, much like the potential within each interaction with ChatGPT. This chapter invites you to step into the role of an artist in the digital realm, using your questions as brushes and ChatGPT as your canvas. The goal? To refine your questioning skills and transform them into precise and impactful prompts that elicit the most insightful responses from AI. Just as an artist refines their technique through practice, you, too, can hone your ability to craft questions that unlock powerful AI interactions.

5.1 Interactive Exercises: Practicing Effective Questioning

Scenario-based exercises are one of the most effective ways to sharpen your questioning skills. Picture yourself in various roles: a customer service representative handling inquiries or a teacher guiding students through complex topics. These scenarios are designed to mimic real-world situations, providing a practical framework for applying your questioning techniques. For customer service, imagine you need to assist a client with a product issue. How would you phrase your question to ChatGPT to ensure you get a detailed and relevant response? Consider how you might prompt ChatGPT to explain a challenging concept to a student in an educational setting. These exercises push you to think critically about structuring your queries for maximum effectiveness, helping you become more adept at engaging with AI.

Role-playing exercises offer another dynamic way to engage with ChatGPT and refine your questioning skills. By adopting different personas, you can explore how varying the tone, style, and context of your questions affects the AI's responses. For instance, you might assume the role of a curious learner, asking open-ended questions to explore a new topic. Alternatively, take on the persona of a business analyst seeking specific data insights. This exercise enhances your adaptability and highlights the importance of context and tone in crafting questions. By experiencing these different perspectives, you gain a deeper understanding of how to tailor your prompts to suit diverse scenarios and objectives.

Feedback mechanisms are crucial in evaluating and improving your question-formulation skills. One valuable tool is reflective journaling. After each exercise, take a moment to jot down your thoughts and observations. What worked well? Where did the AI's response fall short? Reflecting on these interactions allows you to identify patterns and areas for improvement. Peer review sessions can also provide fresh insights. Share your prompts with peers and invite them to offer feedback. Their perspectives can highlight nuances you might have missed, offering new angles to consider in your questioning approach. This self-assessment process is much like a mirror, reflecting your strengths and areas for growth and guiding you toward more effective interactions with ChatGPT.

To keep these exercises both challenging and engaging, it's crucial to offer progressive difficulty levels. Start with beginner-level prompts, focusing on simple, straightforward questions. As you become more comfortable, graduate to intermediate exercises that require more nuance and specificity. Finally, tackle advanced scenarios that demand a high level of precision and creativity in your questioning. This gradual progression mirrors the learning curve of any new skill, allowing you to build confidence and competence at a comfortable pace. Adaptive challenges tailored to your past performance keep the exercises fresh and engaging, pushing you to refine and elevate your questioning techniques continually.

Interactive Element: Question Crafting scenarios.

- **Scenario-Based Prompts:** Write out scenarios and your initial questions.
- **Role-Playing Reflections:** Note different personas and how they influenced your prompts.
- **Feedback Journal:** Document peer feedback and personal reflections after each exercise.
- **Progressive Challenges:** Record your experiences with beginner, intermediate, and advanced prompts.

Keep track of your responses as a tangible record of your progress, offering a structured way to refine and develop your questioning skills. As you document your journey, you'll witness your growth—from basic inquiries to engaging in more sophisticated and meaningful dialogues with ChatGPT.

5.2 Feedback Loop Model: Improving AI Interactions

Imagine having a conversation where each response not only satisfies your query but also helps you craft even better questions the next time around. This is the essence of a feedback loop in AI interactions. It's like having a dance partner who becomes more in sync with every spin and twirl, learning from each misstep and triumph to create a more harmonious dance. In AI, a feedback loop involves a cycle of asking, receiving, and refining. You start with an initial question, analyze the AI's response, and then adjust your follow-up questions based on what you learn. This iterative refinement allows you to hone your querying skills and improve the quality of the AI's responses over time. By continuously engaging in this process, you enhance your understanding of how to interact with AI and help the AI model learn and adapt, making it even more effective in meeting your needs.

Use AI-generated feedback to adjust your prompts to establish an effective feedback loop. Like a chef tweaks a recipe based on taste tests, you can refine your questions based on the AI's answers. If the response is vague, try narrowing your query; if it's too detailed, consider simplifying it. Collecting user input is another key method for refining AI training. Think of it as gathering insights from a focus group—each piece of feedback helps you understand what works and what doesn't. By systematically analyzing the AI's responses, you can identify patterns in how your questions affect the outcomes. This analysis lets you fine-tune your approach, ensuring your interactions become more precise and productive with each iteration.

The role of feedback in learning and adaptation is crucial. Constructive feedback acts as a mirror, reflecting the strengths and weaknesses of your approach. As you identify patterns in user queries, you gain insights into common pitfalls and effective strategies. This knowledge enables you to adjust AI models based on feedback data, much like a gardener pruning a plant to encourage healthy growth. Incorporating feedback into your process creates a dynamic environment where continuous learning and improvement are ongoing. This adaptability ensures that you and the AI develop a more nuanced understanding of engaging effectively, leading to more meaningful and satisfying interactions.

Consider a company implementing a feedback loop to enhance its AI-driven customer service. Initially, they faced challenges with response accuracy and relevance. Collecting customer feedback and AI-generated interactions, they identified key areas for improvement. They improved customer satisfaction and streamlined their service processes through iterative adjustments to their AI models and prompt strategies. Similarly, an educational platform used feedback loops to tailor content to students' needs. By analyzing student interactions, they refined their instructional materials, resulting in more personalized and compelling learning experiences. These examples highlight the transformative power of feedback loops in optimizing AI interactions across various contexts.

Through thoughtful application of feedback loops, you can transform your interactions with AI from static exchanges into dynamic dialogues that evolve and improve over time. By embracing this iterative process, you enhance your ability to engage with AI and contribute to its ongoing development and refinement. The beauty of feedback loops lies in their ability to adapt and grow, creating a symbiotic relationship between you and the AI that benefits both parties. As you continue to explore the potential of AI, remember that each interaction is an opportunity for learning and improvement, driven by the insights and adjustments that feedback loops provide.

5.3 Joining the AI Community: Learning and Support

Picture yourself in a room full of people with common interests, their conversations buzzing with excitement and curiosity. That's what joining an AI community feels like—you're stepping into a vibrant space filled with diverse perspectives and a wealth of ideas. Engaging with others who are passionate about AI not only broadens your understanding but also inspires you to explore new angles and possibilities. These communities are like melting pots where ideas simmer and blend, often leading to innovative breakthroughs. They offer a platform for collaborative learning, where you can gain insights and share experiences, enriching your journey in the AI landscape.

Finding and joining AI communities might initially feel daunting, but it's simpler than you think. Start by exploring online forums and discussion boards dedicated to AI topics. With its various AI subreddits or specialized forums such as AI Stack Exchange, platforms like Reddit offer a treasure trove of discussions and resources. Social media groups on Facebook or LinkedIn also host active AI discussions, bringing together enthusiasts, professionals, and novices alike. These online spaces are accessible and welcoming, making them excellent starting points for anyone eager to dive deeper into AI. By participating in these communities, you connect with like-minded individuals, expanding your network and learning from a global pool of expertise.

Mentorship is pivotal in AI learning, much like a guiding star that helps you navigate complex waters. Connecting with experienced AI professionals can accelerate your skills development, offering guidance and insights that textbooks and online courses might not provide. Mentorship programs are available in many AI communities, allowing you to learn directly from those who have walked the path before you. Whether through formal programs or informal connections, having a mentor means gaining access to invaluable advice, industry knowledge, and career guidance. This relationship can be transformative, helping you avoid common pitfalls and fast-tracking your progress in the AI field.

Consider the story of a newcomer who joined an AI community with little more than curiosity and a willingness to learn. Through active participation and engagement with mentors, they quickly advanced from beginners to AI experts. They honed their skills and contributed to the community's growth by collaborating on projects and sharing their newfound knowledge. Another example is a group of individuals who met through an AI forum and decided to work together on a project. Their collaboration resulted in a successful AI-driven application that gained attention and praise from industry leaders. These stories illustrate the power of community-driven success and highlight the potential that comes with being part of a supportive network.

Textual Element: Community Engagement Checklist

- **Identify Interest Areas:** Determine which AI topics excite you most.
- **Explore Platforms:** Check out forums like Reddit, AI Stack Exchange, and relevant social media groups.

- **Seek Mentorship:** Reach out to experienced members for guidance and support.
- **Participate Actively:** Engage in discussions, share insights, and collaborate on projects.
- **Reflect on Growth:** Regularly assess how community involvement has impacted your learning.

This checklist serves as a practical guide to help you navigate and make the most of AI communities. By following these steps, you can immerse yourself in a world of learning, support, and collaboration, unlocking new opportunities and insights along the way. As you engage with these communities, remember that exchanging ideas is a two-way street—what you contribute is just as valuable as what you gain.

5.4 Online Platforms and Tools for Continuous Learning

In the ever-evolving world of AI, staying updated and informed is crucial. Fortunately, there is a plethora of online platforms that can help you continue learning and growing your skills. AI-focused MOOCs (Massive Open Online Courses) and online courses are fantastic starting points. Websites like Coursera, edX, and Udacity offer comprehensive courses on basic AI concepts and advanced machine learning techniques. These courses are often created in collaboration with top universities and industry experts, providing high-quality content that can help you expand your understanding of AI. Platforms like Khan Academy and FutureLearn offer free resources, allowing you to explore topics at your own pace without financial commitment.

If you're someone who prefers more visual or auditory learning, educational YouTube channels and podcasts are excellent alternatives. Channels like "Two Minute Papers" and "Siraj Raval" break down complex AI topics into digestible segments, making them accessible to beginners and seasoned learners alike. Podcasts such as "AI Alignment Podcast" or "Machine Learning Guide" provide insights into the latest industry trends and developments, allowing you to learn on the go. These resources offer the flexibility to learn wherever you are, whether during a morning commute or while relaxing at home.

One of the most significant advantages of digital learning tools is their accessibility and flexibility. With self-paced learning opportunities, you can tailor your education to fit your schedule. Whether you have a spare fifteen minutes or an entire afternoon, you can dive into a topic that piques your interest without feeling rushed. Interactive tutorials and simulations further enhance this flexibility by offering hands-on practice, allowing you to experiment with AI concepts in a controlled environment. This kind of engagement is invaluable for solidifying your understanding and applying theoretical knowledge to real-world scenarios.

For those looking to put theory into practice, there are platforms specifically designed to facilitate hands-on AI interactions. Virtual AI labs and sandbox environments provide safe spaces to experiment with AI models and algorithms. Platforms like TensorFlow Playground and Google's Teachable Machine let you tinker with machine learning models without needing extensive coding experience. Additionally, AI-powered chatbots such as Replika or ChatGPT offer opportunities to test your questioning and interaction skills, helping you refine your approach and learn from the AI's responses. These tools provide practical experience that complements theoretical learning, bridging the gap between knowledge and application.

Selecting the right learning resources can be daunting, given the vast array of options available. When evaluating course content and quality, consider the instructors' credentials and the depth of the material covered. Look for courses that offer a balanced mix of theory and practical application, ensuring you comprehensively understand the subject. Aligning resources with your skill development needs is also crucial. Identify your expertise level and learning goals (The Four Stages), then choose appropriate resources for your experience and aspirations. This approach ensures that your learning journey is both challenging and rewarding, providing the growth and development you seek.

In this digital age, the tools for learning AI are more accessible than ever, offering a wealth of opportunities to enhance your skills and knowledge. By leveraging these platforms and tools, you can stay informed, engaged, and prepared to tackle the exciting challenges and innovations that AI presents.

5.5 Experimentation as a Learning Tool

Imagine approaching AI like a scientist in a lab, always curious and ready to test out new ideas. Experimentation is not just a method; it's a mindset that can propel your understanding and mastery of AI to new heights. By embracing experimentation, you step into a world of exploration where trying new approaches to problem-solving can reveal unexpected insights. It's about being bold enough to test the boundaries of what AI can do, whether you're tweaking prompts to see how ChatGPT reacts or challenging it with complex scenarios that mimic real-life problems. Regardless of the outcome, each attempt adds another layer to your knowledge and sharpens your skills, turning experimentation into a powerful learning tool.

Experimentation is the birthplace of innovation. It's where creativity meets curiosity, and new ideas are born. When you allow yourself the freedom to experiment, you open the door to discovering novel AI applications that you might not have considered before. Perhaps you uncover a unique way to use ChatGPT in an art project, transforming simple prompts into a collaborative AI-driven installation. Or maybe you stumbled upon a method to optimize a business process through AI trials, leading to more efficient operations and cost savings. These breakthroughs often come from iterative design, where each cycle of testing and refining brings you closer to an effective and innovative solution. This continuous process of trial and error encourages you to view each experiment not as a potential failure but as a stepping stone toward groundbreaking discoveries.

To conduct effective experiments, it's essential to set clear objectives and hypotheses. Think of these as your guiding stars, helping you navigate the vast possibilities of AI exploration. Before you begin, define what you hope to achieve and the questions you aim to answer. This clarity ensures that your experiments are focused and purposeful, increasing the likelihood of meaningful results. Documenting and analyzing the outcomes is equally crucial. Keep a detailed record of each experiment, noting what worked, what didn't, and any surprises along the way. This documentation serves as a valuable reference for future experiments and helps you identify patterns and insights that can inform your next steps. Analyzing the data allows you to refine your approach, ensuring that each experiment builds on the last, driving you closer to your goals.

Consider the story of an artist who used AI to create an interactive installation. They crafted a piece that responded to audience input by experimenting with different prompts and algorithms, creating a dynamic and ever-changing artwork. This innovation captivated viewers, showcasing the potential of AI in creative expression. In another instance, a company conducted AI trials to streamline logistics operations. By testing various AI-driven solutions, they discovered a method that reduced delivery times and costs, significantly boosting their bottom line. These examples highlight the transformative power of experimentation, illustrating how it can lead to remarkable advancements across diverse fields.

Some key takeaways:

- **Promote experimentation as essential for AI mastery**: Explore and experiment with AI tools. Trying new approaches to problem-solving, testing AI capabilities in various scenarios
- **The role of experimentation in innovation**: Experimentation leads to creative breakthroughs. Discovering novel AI applications, iterative design in AI-driven projects
- **Create guidelines for conducting practical experiments**: Promote strategies to ensure productive experimentation. Setting clear objectives and hypotheses, documenting and analyzing experiment outcomes
- **Share examples of successful experimentation:** Present stories of innovative AI applications resulting from experimentation. AI-driven art installations, business process optimizations through AI trials

Experimentation is not just a tool for learning; it's a catalyst for growth and innovation. By integrating experimentation into your approach to AI, you empower yourself to uncover new possibilities, challenge existing assumptions, and drive continuous improvement. So, embrace the spirit of experimentation and let your curiosity lead the way.

5.6 Incorporating AI into Daily Routines

Imagine waking up in the morning and having your coffee brewed, the lights gently turning on, and your schedule for the day laid out—all without lifting a finger. This is the magic of integrating AI into daily life. By automating household tasks, AI can significantly enhance your productivity. Smart home devices like automated thermostats and lighting systems adjust settings based on your preferences, creating a comfortable environment effortlessly. AI-powered vacuum cleaners can keep your floors spotless while you focus on more meaningful activities. These technologies save time and reduce the mental load of daily chores, freeing you to engage in activities you truly enjoy.

Beyond home automation, AI can be a personal assistant that keeps you organized. Imagine never missing an appointment because your AI assistant sends timely reminders and adjusts your schedule as needed. Using AI for task management simplifies planning, allowing you to prioritize tasks and allocate time more efficiently. It's like having a personal secretary who works around the clock, ensuring that nothing slips through the cracks. AI can also offer personalized suggestions for leisure activities, like recommending books based on your reading habits or suggesting new recipes to try. These minor enhancements add up, making your life smoother and more enjoyable.

Integrating AI into your routine may seem daunting, but it can be surprisingly seamless. Start by setting up AI-powered smart home devices. For example, many smart speakers and hubs can control various gadgets in your home from a single interface. This centralization makes managing everything from lighting to security systems easy with just a few voice commands. Additionally, AI assistants like Siri, Alexa, or Google Assistant can be utilized for daily task management. These tools can help set reminders, make shopping lists, and even answer trivia questions, reducing the need for multiple apps or devices. By taking these small steps, you can gradually incorporate AI into your life, enhancing your efficiency without overwhelming yourself.

The long-term impact of using AI daily is profound. As you continue to interact with AI, you naturally develop skills in navigating and optimizing these technologies. This ongoing use fosters continuous learning and adaptation as you discover new features and applications that further enhance your lifestyle. Over time, you build AI literacy, becoming more comfortable and adept at leveraging technology to meet your needs. This growth not only improves your personal efficiency but also prepares you for an increasingly tech-driven world where AI skills are becoming essential. Embracing AI in everyday life is a proactive step toward personal development, equipping you with the tools needed to thrive in the modern age.

Consider the story of a family who transformed their home with AI technology. By integrating smart thermostats, lighting, and security systems, they created a home environment that adjusted to their needs automatically. This automation enhanced their comfort and significantly reduced their utility bills. On a professional level, imagine a busy manager who uses AI for time management. By relying on an AI assistant to schedule meetings and prioritize tasks, they increased their productivity and reduced stress, allowing more time to focus on strategic planning and team development. These examples demonstrate the tangible benefits of AI integration, illustrating how it can elevate both personal and professional aspects of life.

Weaving AI into your daily routines opens the door to a more efficient and enriched lifestyle. The benefits extend beyond convenience, fostering growth and development that align with the demands of today's fast-paced world. As you explore these possibilities, remember that each small step in integrating AI brings you closer to a future where technology seamlessly supports and enhances your everyday experiences.

Chapter Six

Advanced Prompt Strategies

Think back to a time when you found yourself staring at a whiteboard, marker in hand, trying to untangle the complexities of a major project. You might have felt like an artist facing a blank canvas, unsure of where to begin. This is where ChatGPT steps in, transforming that daunting blankness into a vibrant brainstorming session. Imagine having the power to break down complex problems into manageable pieces, each one a brushstroke that brings clarity to your grand vision. This chapter will guide you on leveraging ChatGPT to tackle intricate challenges across various fields, from strategic business planning to technical troubleshooting in IT.

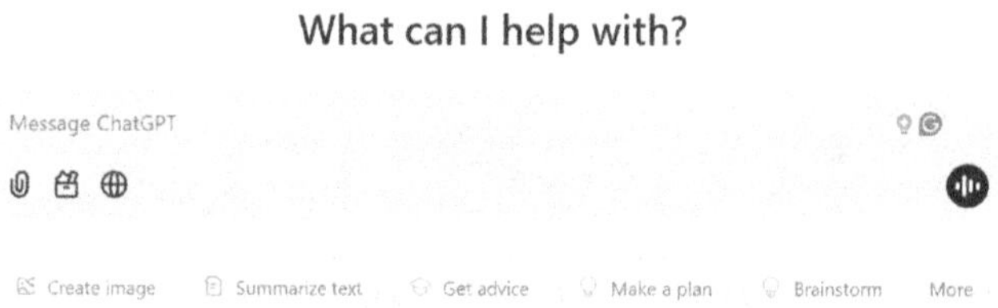

Using AI for complex problem-solving might sound like something out of a sci-fi movie, but it's very much a reality today. Think of strategic business planning, where the stakes are high, and decisions need to be data-driven and forward-thinking. ChatGPT can analyze trends, suggest innovative strategies, and even simulate various scenarios to help you make informed decisions. It's like having a strategic advisor who never tires, offering insights that might take a human team days to uncover. Whether exploring new markets or refining existing operations, ChatGPT provides a fresh perspective, highlighting opportunities and pitfalls you might not have considered.

In the realm of IT, ChatGPT shines as a technical troubleshooter. Imagine a server going down in the middle of a busy workday. ChatGPT can help diagnose the issue, suggesting steps to resolve it quickly. Analyzing error messages and system logs provides insights that guide IT professionals through the troubleshooting process, minimizing downtime and maintaining productivity. This capability is invaluable in environments where time is of the essence and every minute counts. ChatGPT is a knowledgeable partner, helping IT teams navigate complex technical challenges quickly and efficiently.

Framing complex problems requires a structured approach to prompt design. Think of it as building a puzzle. Each piece represents a component of the problem, forming a complete picture together. Start by breaking down the problem into smaller, manageable parts. This decomposition makes it easier for ChatGPT to process and respond effectively. For instance, if you're developing a new product, identify each stage of the process, from market research to prototype testing. Using hypothetical scenarios to explore different outcomes allows ChatGPT to suggest creative solutions or highlight potential hurdles. This method transforms abstract challenges into clear, actionable steps that guide you toward a solution.

Real-world examples illustrate the power of AI in solving complex problems. Consider the challenge of supply chain disruptions, a common issue in today's globalized economy. ChatGPT can analyze data from multiple sources, identifying patterns and suggesting strategies to mitigate delays. Simulating different scenarios helps businesses prepare for potential disruptions, ensuring they have contingency plans in place. Similarly, companies face unique challenges when entering emerging markets that require tailored strategies. ChatGPT can assess market trends, consumer behavior, and competitor analysis, offering insights that guide businesses in crafting effective entry strategies. These examples highlight how AI can transform challenges into opportunities, providing a competitive edge in dynamic environments.

Visual Element: AI in Problem-Solving Infographic

- **Strategic Business Planning:** AI analyzes trends, simulates scenarios, and suggests strategies.
- **IT Troubleshooting:** Diagnoses issues, offers solutions, and minimizes downtime.
- **Supply Chain Solutions:** Identifies patterns, mitigates disruptions, and ensures continuity.
- **Market Entry Strategies:** Assesses trends, analyzes competitors, and guides strategic decisions.

These visual elements demonstrate ChatGPT's capability to address complex challenges across diverse fields, empowering you to leverage AI for more intelligent, effective problem-solving.

Case studies further illustrate AI's impact on complex problem-solving. In environmental impact assessments, AI analyzes vast amounts of data to identify potential risks and suggest mitigation strategies. This capability is crucial for industries like construction and energy, where understanding environmental impact is key to sustainable development. In healthcare, AI assists in medical diagnostics and treatment planning by analyzing patient data and recommending treatment options. This enhances the accuracy of diagnoses and supports healthcare professionals in delivering personalized care. These case studies underscore AI's transformative power, showcasing how ChatGPT and similar tools can drive innovation and efficiency in tackling complex challenges.

6.1 Multi-Step Prompts for Detailed Responses

Imagine you're piecing together a jigsaw puzzle. Each piece is crucial and must fit perfectly with the others to reveal the whole picture. Multi-step prompting with ChatGPT works much the same way. By breaking down a query into sequential steps, you guide the AI through a journey of discovery, allowing it to produce comprehensive and nuanced responses. Picture this as a step-by-step analysis where each question builds upon the last. This layered questioning method uncovers deeper insights, akin to peeling back the layers of an onion to get to the core. It's an approach that transforms complex topics into manageable segments, making navigating intricate subjects easier for you and the AI.

To construct effective multi-step prompts, establish a logical flow between your questions. Think of it as crafting a storyline, where each question leads naturally to the next, maintaining a clear and coherent narrative. This continuity keeps the AI-focused and ensures that responses are relevant throughout the process. **Clarity** is key here; each question should be **straightforward** and **unambiguous**, guiding the AI without overwhelming it. Imagine conducting a detailed SWOT analysis. Begin by asking, "**What** are the strengths of our business?" Follow this with, "**What** weaknesses should we address?" By structuring your prompts in this way, you encourage the AI to explore each aspect thoroughly before moving on, resulting in a well-rounded analysis. This is also a process known as the **5W's**, which we will discuss in more detail in the bonus content section.

Consider the challenges you might face with multi-step prompting. One common issue is managing context retention across steps. The AI might lose track of previous interactions, leading to disjointed responses. To overcome this, provide context at the beginning of each new prompt, reminding the AI of the ongoing narrative. This practice helps maintain the conversation thread, ensuring that each response builds logically on the last. Another challenge is keeping the user engaged throughout the process. Long sequences of questions can become tedious, so it's essential to vary the type and style of your prompts. Mix factual questions with creative ones or introduce hypothetical scenarios to keep the interaction dynamic and engaging.

Take the example of developing project plans with phased implementation. You might start with, "What is the first step in our project timeline?" Once the AI provides a response, follow up by asking, "How can we ensure the success of this phase?" This sequence not only breaks the project down into manageable parts but also encourages the AI to delve deeper into each phase, offering insights and suggestions that might not surface with a single prompt. Using multi-step prompts effectively transforms your interaction with ChatGPT into a collaborative process, where each step is a building block toward a comprehensive and insightful outcome.

Interactive Element: Multi-Step Prompt Exercise

Take a complex topic you're working on, such as a new marketing strategy, and create a series of prompts. Start with a broad question, "What are our target demographics?" Use the AI's response to guide your next question: "How can we tailor our messaging to this audience?" Continue this process, crafting a series of interconnected prompts that lead to a detailed and actionable strategy. Reflect on how each step builds on the previous one and adjust your prompts for clarity and continuity.

6.2 Adapting Prompts for Different Contexts

Consider context adaptation's importance when crafting prompts for AI like ChatGPT. It's similar to tailoring a suit; one size certainly doesn't fit all. In the realm of AI interactions, adapting prompts to specific scenarios enhances their relevance and effectiveness. When you align prompts with industry-specific needs, you're essentially speaking the language of your field, whether it's legal, medical, or marketing. This alignment ensures that the AI understands the nuances and priorities unique to your sector, providing responses that are not only accurate but also insightful. Similarly, tailoring questions to accommodate cultural nuances is crucial, especially in our globalized world. Using the correct cultural references or idioms can significantly affect how AI interprets and responds to prompts.

So, how can you effectively adapt your prompts to suit different contexts? Start by researching the contextual factors before you even begin. This might involve understanding a specific industry's regulations or a particular region's cultural expectations. By arming yourself with this knowledge, you're better equipped to craft prompts that resonate. Using localized language and examples is another powerful strategy. It's like speaking directly to your audience, making them feel seen and understood. For instance, when constructing a prompt for a marketing campaign in Japan, incorporating familiar cultural symbols and language nuances can enhance the relevance and impact of the AI's responses.

Let's explore some examples where contextually adapted prompts led to successful outcomes. The same question might yield different answers in legal inquiries across different jurisdictions due to variations in laws and regulations. By adjusting your prompts to reflect these differences, ChatGPT can provide guidance that's not just legally sound but also tailored to the specific jurisdiction you're dealing with. Similarly, context adaptation can be the key to success when developing marketing strategies for regional markets. A campaign that works wonders in one country might flop in another if cultural differences aren't taken into account. By crafting prompts considering regional preferences and behavior, you can ensure that the AI's suggestions are relevant and effective.

AI's ability to understand and apply contextual elements is a game-changer. Through AI-driven sentiment analysis, ChatGPT can gauge the emotional tone of a conversation, allowing it to tailor its responses accordingly. This capability is invaluable in customer service scenarios, where understanding a customer's mood can significantly impact the interaction's outcome. Language translation and localization support are other areas where AI excels. By translating prompts into the local language and adjusting for cultural nuances, ChatGPT can engage more effectively with users from diverse backgrounds. This adaptability enhances communication and ensures that the AI's responses are both culturally sensitive and contextually appropriate.

As you explore these strategies, remember that the goal is to enhance the relevance and impact of AI interactions. By adapting your prompts to suit different contexts, you empower ChatGPT to provide responses that are not only accurate but also resonate with the unique needs and expectations of your audience.

6.3 Utilizing AI for Data-Driven Decision Making

Imagine standing in a control room, surrounded by monitors displaying streams of data from every corner of your business. Each screen holds the potential to unlock insights that could transform your strategies and boost your success. This is where ChatGPT comes into play, acting as your data interpreter and guide. In our data-rich world, making informed decisions is no longer just a competitive advantage—it's a necessity. ChatGPT can sift through mountains of data, highlighting trends and patterns that might escape the human eye. Imagine analyzing data trends to predict future business opportunities, much like a meteorologist forecasts the weather. With ChatGPT, you can tap into its computational prowess to decipher complex data sets and generate forecasts that guide strategic planning and resource allocation.

Customer behavior can often feel like a puzzle, with pieces scattered across different data points and touchstones. ChatGPT excels in identifying these patterns, piecing together the puzzle to reveal insights that inform marketing strategies and product development. By analyzing purchase histories, browsing behaviors, and even social media interactions, ChatGPT helps you understand what drives your customers' decisions. This understanding allows you to tailor your offerings and communications, ensuring you effectively meet their needs and expectations. The power of AI lies in its ability to connect the dots, offering a comprehensive view that enables you to anticipate trends and stay ahead of the competition.

Incorporating data into prompts is a strategic move that enhances the depth and accuracy of AI responses. Think of it as seasoning a dish to perfection—just the right amount of data can transform a bland response into something insightful and actionable. When crafting prompts, consider integrating statistical data for predictive modeling. This approach allows ChatGPT to leverage historical data and current trends to offer evidence-based predictions and recommendations. Additionally, presenting visual data representations, such as graphs or charts, can clarify complex concepts, making it easier for the AI to process and respond effectively. These visual aids act as a bridge, translating raw data into a format that enhances understanding and decision-making.

Consider the world of marketing, where optimizing campaigns through data insights can be the difference between thriving and merely surviving. ChatGPT can analyze campaign performance, identifying which strategies resonate with your audience and which fall flat. By understanding these dynamics, you can refine your approach, ensuring that your marketing efforts are both impactful and cost-effective. In financial planning, AI's ability to analyze historical data provides a solid foundation for making investment decisions, budgets, and growth strategies. ChatGPT offers insights that support sound financial planning by evaluating past performance and market conditions, positioning your organization for long-term success.

Case studies abound, showcasing organizations successfully leveraging AI for strategic decision-making. In retail, companies use AI to optimize inventory management, ensuring that shelves are stocked with the right products at the right time. By analyzing sales data, seasonal trends, and customer preferences, AI helps retailers anticipate demand and minimize waste. This precision not only improves operational efficiency but also enhances customer satisfaction. Healthcare providers, too, are harnessing AI's potential and using it to optimize patient care. By analyzing patient histories, treatment outcomes, and medical research, AI supports healthcare professionals in making data-driven decisions that improve patient outcomes. This application highlights AI's transformative power, offering a glimpse into how data-driven decision-making can enhance business performance and quality of life.

6.4 Personalizing AI Interactions for Greater Relevance

Imagine walking into your favorite coffee shop, where the barista knows your name and your go-to order. Personalization in AI aims to replicate that feeling of being recognized and understood. When AI interactions are tailored to your specific needs and preferences, the experience becomes more meaningful and satisfying. Personalized AI can recommend books that match your taste, suggest workout routines that align with your fitness goals, or even curate playlists that lift your mood. This customization enhances user satisfaction because it feels like the AI genuinely understands you, leading to more effective and enjoyable interactions.

Personalizing AI interactions involves incorporating user-specific information into prompts, making the responses more relevant. One effective strategy is to use user data to inform the content of prompts. For instance, if you're using an AI assistant for fitness coaching, it could consider your workout history, health goals, and even dietary preferences to suggest personalized exercise plans. This method ensures that the AI's guidance is aligned with your objectives and tailored to your lifestyle. Another approach is adjusting the tone and style of interactions based on user preferences. Some people prefer concise, to-the-point advice, while others enjoy a more conversational exchange. By adapting the AI's communication style, you create an experience that resonates with your personal taste, making each interaction more engaging.

Consider the scenario of personalized fitness coaching through AI. Imagine receiving workout recommendations that are not just generic but specifically designed around your fitness level, goals, and even the equipment you have at home. This level of personalization keeps you motivated and on track, as the AI provides advice that fits seamlessly into your routine. Similarly, when planning a vacation, AI can craft custom travel itineraries based on your interests, preferred travel pace, and past vacations. Whether you're a history buff, an adventure seeker, or a culinary enthusiast, the AI tailors its suggestions to ensure you have a memorable experience. These examples highlight how personalization can significantly enhance user engagement, making interactions with AI more relevant and impactful.

The tools and technologies enabling AI personalization are constantly evolving. Customer Relationship Management (CRM) systems are crucial in managing user data, allowing AI to access and analyze information that informs its responses. By integrating CRM with AI, businesses can create personalized experiences that strengthen customer relationships and increase satisfaction. Furthermore, AI algorithms designed for adaptive learning continuously refine their understanding of user preferences, ensuring that the personalization remains relevant over time. These algorithms learn from previous interactions, adjusting their recommendations and communication style to better align with user needs. This adaptability is key to maintaining a personalized experience that evolves alongside the user.

Platforms like Netflix and Spotify have mastered the art of AI personalization, using sophisticated algorithms to deliver content that resonates with individual users. They provide tailor-made recommendations by analyzing viewing habits and listening patterns, enhancing user engagement and loyalty. In the educational sphere, personalized learning paths are becoming increasingly popular. AI can assess a student's learning style, strengths, and areas for improvement to create a customized curriculum that maximizes their potential. This approach improves educational outcomes and fosters a love for learning by making the experience more enjoyable and rewarding. As AI advances, the possibilities for personalization are boundless, offering new opportunities to create interactions as unique and varied as the users themselves.

6.5 Advanced Techniques for Creative Writing and Content Generation

Imagine sitting at your desk, a blank page staring back at you, waiting for the next great idea to emerge. This is where ChatGPT steps in, turning that blank page into a canvas of possibilities. With its boundless capacity for processing language, AI has become a remarkable ally in creative endeavors, offering a fresh perspective on generating original content. Whether you're developing a story plot, crafting a character arc, or brainstorming marketing copy, ChatGPT can provide the spark you need. It acts as a creative partner, offering suggestions that might not have crossed your mind and nudging you toward new directions and innovations. By providing a framework or even just a single idea, ChatGPT can help you overcome creative blocks, making the writing process more productive and enjoyable.

To enhance creativity with AI, consider using prompts as a tool for brainstorming sessions. Think of them as conversation starters that can lead to unexpected insights. Present a scenario or a question to ChatGPT and let the AI's responses guide your thoughts. This method encourages divergent thinking, where the goal is to explore many possibilities rather than focus on a single solution. By exploring these different paths, you open yourself up to a wealth of ideas, some of which might be the seeds of your next great project. AI exploration can also push you to think outside the box, challenging conventional norms and encouraging you to view problems from new angles. This approach not only enriches your creative process but also broadens your perspective, making your work more dynamic and engaging.

AI-assisted creative projects are becoming increasingly common, showcasing the potential of technology in artistic fields. Collaborative writing projects, for instance, benefit significantly from AI input. Imagine working with a team of writers, each bringing their unique voice to the table. ChatGPT can act as an additional team member, offering suggestions for plot twists or character developments that enhance the narrative. This collaborative process is akin to a jam session, where each participant contributes to the piece, creating a richer and more nuanced final product. Similarly, AI-generated artwork and music composition are gaining traction, with AI providing inspiration and even technical execution for artists and musicians. These projects highlight the synergy between human creativity and AI, producing innovative and captivating works.

Balancing AI and human creativity is crucial for maintaining the integrity of artistic expression. AI is a tool for inspiration and ideation, providing a starting point or a new perspective that can enhance human creativity. It's important to remember that while AI can generate ideas, the human touch brings depth and emotion to the work. Artistic expression is inherently personal, reflecting the creator's experiences, feelings, and worldview. AI can offer guidance and support, but the human element infuses the work with meaning and resonance. By collaborating with AI, artists can expand their creative horizons, exploring new possibilities while preserving the authenticity of their voices.

In the broader scheme of creativity, AI doesn't replace human ingenuity but complements it. This partnership opens doors to new forms of expression, blending technology with traditional artistry. As you explore the potential of AI in your creative projects, consider how this collaboration can enhance your work, making it more diverse and impactful. With AI as your creative ally, the possibilities are limitless, offering a canvas where your imagination can soar.

As we wrap up this chapter, remember that AI is a tool to enhance your creative endeavors, not replace them. In the next chapter, we will explore how to integrate these advanced techniques into practical applications that can transform your daily interactions with AI. Stay curious and open-minded as we continue to unlock the potential of ChatGPT in everyday scenarios.

Quiz: Are you progressing through the 4 stages of learning and skill acquisition?

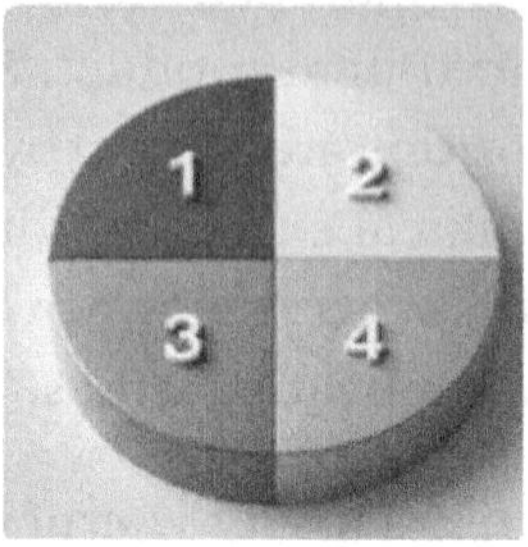

Chapter Seven

Building AI Literacy and Tech Savvy

"AI and machine learning are a big deal. They're going to be a huge enabler for companies like Amazon."
Jeff Bezos (Founder of Amazon)

"AI is the defining technology of our time. It's transforming industries, changing how we work, and driving the next wave of innovation."
Satya Nadella (CEO of Microsoft)

Picture this: you walk into a bustling tech conference, eager to soak up the latest innovations in artificial intelligence. Yet, as conversations buzz around you, you feel like you're missing the secret code everyone else seems to know. This chapter is your key to unlocking that code. It's about demystifying the world of AI and equipping you with the knowledge to engage confidently and effectively. You'll gain a solid grounding in key AI concepts and learn how they apply to real-world scenarios, making you not just a participant in the tech dialogue but an informed contributor.

Understanding artificial intelligence starts with distinguishing between its core components: **artificial intelligence, machine learning, and deep learning**. AI is the overarching concept, encompassing any technique that enables computers to mimic human intelligence. Machine learning, a subset of AI, involves teaching computers to learn from data, improving their performance over time without being explicitly programmed. Within machine learning lies deep learning, which uses neural networks with many layers (hence "deep") to analyze various levels of abstraction. These distinctions are crucial in grasping how AI systems operate and evolve, offering insight into the mechanisms that drive innovations across industries.

We dive into how AI systems are trained when discussing supervised and unsupervised learning. Supervised learning involves teaching AI using labeled datasets—think of it as learning with a teacher. The AI receives inputs and correct outputs, learning to map the two. In unsupervised learning, the AI explores data without explicit instructions, identifying patterns and relationships on its own—imagine learning by exploring a new city without a map. These approaches determine how AI models develop their understanding and capabilities, shaping their applications in tasks like recommendation engines and data clustering.

To see AI in action, consider how neural networks revolutionize image recognition. These networks operate like a brain's neurons, recognizing patterns and features to identify objects within images. It's the technology behind facial recognition in your smartphone or the tagging of friends in social media photos. Natural Language Processing (NLP) powers virtual assistants like Siri and Alexa, enabling them to understand and respond to your spoken requests. NLP dissects language to grasp meaning and context, making interactions feel natural and intuitive. These applications illustrate the practical impact of AI concepts, transforming abstract ideas into tangible, everyday tools.

Looking ahead, quantum computing promises to redefine AI's potential. This emerging technology leverages quantum bits, or qubits, to process information at unprecedented speeds, tackling unsolvable problems for traditional computers. As quantum computing advances, it could significantly accelerate AI development, enhancing predictive modeling, data analysis, and more. It's a frontier of innovation that holds immense promise for the future, offering glimpses of possibilities that stretch beyond our current technological landscape.

For anyone keen to deepen their AI literacy, numerous resources are available. Online courses and certification programs from platforms like **Coursera** and **edX** offer structured learning paths, covering everything from basic AI principles to advanced machine learning techniques. AI-focused publications and journals, such as the Journal of Artificial Intelligence Research, provide in-depth insights into current trends and breakthroughs. Engaging with these resources cultivates a robust understanding of AI, empowering you to navigate and leverage its capabilities effectively.

Interactive Element: AI Literacy Resource List

- **Online Courses:** Explore platforms like Coursera, edX, and Udacity for comprehensive AI courses.
- **Publications and Journals:** Stay updated with the Journal of Artificial Intelligence Research and AI Magazine.
- **Community Engagement:** Join forums and AI groups on social media for discussions and networking.
- **Certifications:** Consider certifications from recognized institutions to validate your AI expertise.

These resources are your stepping stones to becoming AI literate, enabling you to participate actively in the evolving tech landscape.

7.1 Understanding the AI Tech Stack

Imagine building a house. You need a solid foundation, walls, a roof, and all the components that make it livable. An AI tech stack is similar, comprising various layers and tools that construct a functioning AI system. You'll find the hardware at the base of this stack, often in Graphics Processing Units (GPUs). These powerhouses handle the heavy lifting, rapidly processing large volumes of data, which is crucial for training complex AI models. GPUs accelerate the computations needed for deep learning tasks, enabling the AI to learn faster and more efficiently. They're the muscle behind the brain, ensuring your AI is quick and responsive. Moving up the stack, software frameworks like TensorFlow and PyTorch come into play. These frameworks are the architects of AI, providing the building blocks for developing and training models. They offer pre-built modules and libraries that simplify the creation of neural networks and other AI structures, making them accessible even for those new to the field. TensorFlow, for instance, is known for its flexibility and scalability, allowing developers to experiment with different models. On the other hand, PyTorch is praised for its simplicity and dynamic computation, making it a favorite among researchers and hobbyists. These frameworks are the tools that shape the raw computational power of GPUs into intelligent, functional AI systems.

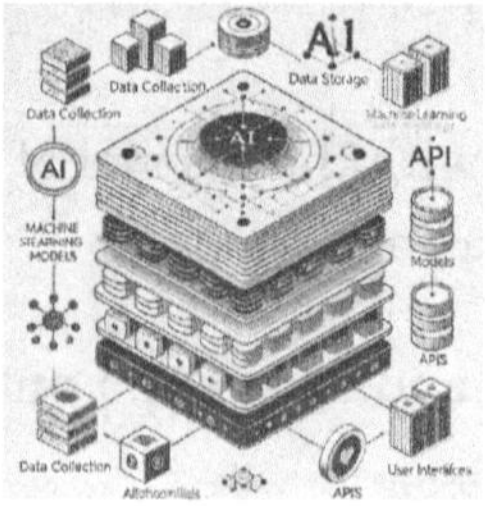

Data storage solutions are another crucial component in managing the vast datasets AI systems rely on. These solutions ensure data is organized, secure, and easily accessible, which is essential for training and refining AI models. Cloud computing takes this a step further, offering scalable AI applications that can expand as needed. With cloud services like AWS and Google Cloud, you can tap into virtually unlimited storage and processing power, removing the constraints of physical hardware. This scalability is a game-changer for AI development, allowing both startups and established companies to innovate without the burden of hefty infrastructure costs.

However, integrating these components isn't always straightforward. Compatibility issues between different software tools can arise, much like trying to fit mismatched puzzle pieces together. Ensuring smooth communication between various parts of the tech stack is crucial for seamless AI operation. Data transfer and synchronization challenges can also pose hurdles, mainly when dealing with large, distributed datasets. Timely and accurate data transfer is vital for maintaining the integrity and performance of AI systems. Overcoming these obstacles requires meticulous planning and often involves employing specialized tools or custom solutions to bridge compatibility gaps.

Take, for example, a tech firm that optimized its AI workflows by refining its tech stack. By integrating efficient data storage with powerful GPUs and leveraging cloud computing, they cut down processing times significantly, allowing them to iterate and innovate faster. Startups, particularly those leveraging cloud-based AI solutions, have found success by using scalable resources that grow with their needs. This approach reduces initial costs and delivers flexibility, enabling them to focus more on developing their AI without worrying about infrastructure limitations. Such strategic use of the AI tech stack has empowered companies to push boundaries and deliver cutting-edge solutions across various industries.

7.2 Exploring AI's Role in Digital Transformation

Imagine walking into a hospital where the doctor, aided by artificial intelligence, can diagnose a patient's condition with remarkable speed and accuracy. This isn't science fiction; it's digital transformation at work. Digital transformation refers to how businesses and other organizations integrate digital technology into all aspects of their operation, fundamentally changing how they operate and deliver value to customers. AI plays a pivotal role in this change by automating manual processes and enhancing the customer experience through personalization. Whether it's through chatbots that provide 24/7 customer service or algorithms that predict what products you might like based on past purchases, AI is reshaping how we interact with technology on a daily basis. By increasing efficiency and streamlining operations, AI empowers businesses to focus more on innovation and less on routine tasks.

Healthcare is one sector that is experiencing a significant shift due to AI. In diagnostics, AI assists doctors by analyzing medical images with precision, spotting anomalies that the human eye could easily miss. This not only speeds up the diagnostic process but also improves accuracy, potentially saving lives. Similarly, AI optimizes inventory management in retail by predicting demand, reducing waste, and ensuring that shelves are always stocked with the right products. This application of AI helps retailers manage their inventories more effectively, minimizing costs and increasing customer satisfaction. These examples illustrate how AI is enhancing current processes and paving the way for entirely new ways of operating that were previously unimaginable.

Despite these advancements, integrating AI into business processes comes with its own set of challenges. On the one hand, AI can significantly boost efficiency and reduce costs. For instance, automated customer service can handle queries quickly and consistently, freeing up human employees for more complex tasks. On the other hand, there are ethical considerations to ponder. The use of AI in decision-making can lead to workforce displacement as machines take over roles traditionally held by people. This raises questions about job security and the future of work. Moreover, ethical concerns about data privacy and the potential for bias in AI algorithms must be addressed to ensure that AI systems are fair and equitable.

Consider financial institutions that have successfully implemented AI for fraud detection. By analyzing transaction patterns and identifying anomalies, AI systems can detect fraudulent activities much faster than human analysts. This not only protects customers but also saves institutions millions in potential losses. In manufacturing, companies use predictive maintenance powered by AI to anticipate equipment failures before they occur. By analyzing data from machinery, AI can predict when maintenance is needed, reducing downtime and optimizing productivity. These case studies highlight how businesses across various sectors are adapting to AI and thriving with it, using these technologies to gain a competitive edge and drive growth.

The journey of digital transformation, powered by AI, is as much about overcoming challenges as it is about leveraging opportunities. As businesses continue integrating AI into their processes, they must balance the benefits of increased efficiency and cost savings with the ethical responsibilities of using such powerful technology. This balance is essential for ensuring that AI serves the interests of businesses and the broader society.

7.3 How Machine Learning Algorithms Work

Imagine teaching a child how to recognize different animals. You start with a few pictures, and over time, the child learns to distinguish a cat from a dog. Machine learning algorithms work in much the same way. They're designed to identify patterns and make decisions based on data. Decision trees, neural networks, and clustering hold significant roles among these algorithms. Decision trees break down data into branches, much like a flowchart, helping to make decisions by following a path of yes/no questions. Neural networks, inspired by the human brain, consist of interconnected nodes that work together to solve complex problems. Clustering, however, groups sets of data points without prior labels, identifying inherent structures within the data.

These algorithms fall into categories based on their learning processes. Supervised learning involves training the model with data that has known outcomes, similar to using a study guide before a test. This method teaches the algorithm to predict outcomes based on input data. Unsupervised learning, however, allows the model to explore the data without guidance, discovering patterns and relationships independently, like finding your way through a new city without a map. Reinforcement learning is another approach where algorithms learn by trial and error, much like training a pet with rewards and consequences. Each type of learning offers unique advantages, allowing algorithms to adapt to various tasks and challenges.

Training machine learning models involves several crucial steps. It all begins with data preprocessing, which is akin to cleaning and organizing ingredients before cooking a meal. This step includes removing irrelevant data, handling missing values, and transforming data into a suitable format. Feature selection follows, where you choose the most essential variables that will influence the model's predictions, much like picking the best ingredients for a recipe. Once the data is prepped, the model is trained to recognize patterns and make predictions. Model evaluation comes next, testing how well the algorithm performs with new data. Hyperparameter tuning is the final step, where adjustments are made to improve the model’s accuracy. This process ensures that the algorithm learns effectively, providing reliable outcomes.

Machine learning algorithms power many AI systems we encounter daily. In streaming services, recommendation engines analyze your viewing history to suggest shows or movies you might enjoy, enhancing your user experience. These algorithms evaluate patterns in your behavior, predicting what you're likely to watch next. In banking, fraud detection models sift through vast amounts of transaction data, identifying unusual patterns that may indicate fraudulent activity. These systems act like vigilant security guards, monitoring transactions and alerting human analysts when something seems off. By leveraging machine learning, these applications can process information at a scale and speed that would be impossible for humans alone, making them invaluable tools in their respective fields.

Developing machine learning models isn't without its challenges. Overfitting occurs when a model learns too much from the training data, capturing noise along with the signal. It's like memorizing answers without understanding the questions, leading to poor performance with new data. Conversely, underfitting happens when a model is too simple, failing to capture the underlying patterns. This results in inaccurate predictions, much like a student who hasn't studied enough for a test. Data bias is another critical concern, as it can skew the algorithm's decisions, potentially leading to unfair or inaccurate outcomes. Ensure that data is diverse and representative, which is crucial to mitigating these biases. Ethical considerations also play a role, as algorithms must be designed with fairness and transparency in mind, avoiding discrimination and ensuring responsible use.

7.4 The Importance of Data in AI Development

Imagine AI as a growing tree. Just as a tree needs sunlight and water to flourish, AI requires data. Data is the foundation for AI, shaping its learning processes and influencing its development. Good data is crucial because it teaches the AI how to interpret the world, make decisions, and improve its accuracy over time. Think of data as the textbooks AI reads. The quality of these textbooks matters immensely. High-quality data ensures the AI learns correctly and efficiently, while poor data can lead it astray, resulting in inaccurate models and unreliable outputs. Without proper data, AI systems would be like students without study materials—ill-prepared and inconsistent.

Data comes in many forms, each serving a unique purpose in AI projects. Structured data resembles a neatly organized spreadsheet, with rows and columns that clearly define relationships between different pieces of information.

This type is easy for AI systems to process and analyze. Unstructured data, on the other hand, is more like a jumble of notes—think text documents, images, and videos—requiring more effort to decipher. Real-time data collection allows AI to analyze information as it happens, enabling quick responses and decisions. This is akin to getting live updates during a sports game, allowing AI to adjust strategies on the fly. The diversity and richness of data types help AI systems learn and adapt to a wide array of scenarios.

Managing this data effectively is no small task. It starts with data cleaning, much like tidying up a workspace. This process involves removing duplicates, correcting errors, and filling in missing values, ensuring only the best data feeds the AI. Preprocessing transforms raw data into a format AI can understand, akin to translating a book into a language the AI speaks fluently. Data security and privacy measures are equally crucial. They protect sensitive information and ensure compliance with regulations, akin to guarding a treasure chest with valuable contents. By implementing robust data management practices, AI developers can create models that are not only accurate but also ethical and trustworthy.

The impact of effective data use is evident in numerous success stories. Consider smart cities, which utilize data to enhance urban planning and improve the quality of life for residents. By analyzing traffic patterns, energy consumption, and public safety data, cities can optimize transportation routes, manage resources efficiently, and respond swiftly to emergencies. This data-driven approach transforms urban areas into responsive and adaptive environments. Precision farming leverages data to maximize crop yields and minimize environmental impact in agriculture. Farmers use soil sensors, weather forecasts, and satellite imagery data to make informed decisions about planting, irrigation, and pest control. This tailored approach enhances productivity while conserving resources, demonstrating the power of data in driving sustainable practices.

7.5 Ethical Considerations in AI Use

In our increasingly digital world, the ethical landscape surrounding AI is becoming more complex and significant. One of the major ethical challenges involves privacy concerns in data collection. As AI systems become more integrated into our daily lives, they gather vast amounts of data—from shopping habits to personal preferences. This data collection raises questions about how much information is too much and who gets to see it. Are we comfortable with companies knowing intimate details of our lives, and more importantly, are they safeguarding that information responsibly? These concerns highlight the importance of establishing clear boundaries to protect user privacy while still allowing AI to function effectively.

Another pressing issue is bias and fairness in AI decision-making. AI systems learn from data, and if that data reflects societal biases, the AI will likely replicate them. This can lead to unfair outcomes, like biased hiring practices or discriminatory loan approvals. Ensuring fairness means recognizing these biases and actively working to minimize them. It's about creating AI that reflects the best of human decision-making, not its flaws. Addressing these biases requires a concerted effort to scrutinize data sources, diversify training datasets, and implement checks that ensure equitable outcomes.

The importance of responsible AI practices cannot be overstated. Transparency in AI algorithms and processes is crucial. Users and stakeholders must understand how decisions are made and on what basis. This transparency builds trust and accountability, clarifying that AI isn't a mysterious black box but a tool governed by understandable principles. Ensuring accountability and oversight involves setting up frameworks that monitor AI systems for performance and ethical compliance. It's about creating a culture where AI development is guided by clear moral standards, with checks in place to address any deviation.

Real-world scenarios often bring these ethical dilemmas into sharp focus. Take autonomous vehicles, for example. In emergencies, these vehicles must make split-second decisions that could have life-or-death consequences. How do we program a car to choose between hitting a pedestrian or swerving into oncoming traffic? These scenarios highlight the moral complexities AI developers face. Similarly, the use of AI in surveillance raises societal concerns. While AI can enhance security, its potential for misuse—such as infringing on personal freedoms—cannot be ignored. Balancing security with privacy rights is a delicate act that requires careful ethical consideration.

Promoting responsible AI use involves developing and adhering to ethical guidelines for AI projects. These guidelines should outline the values and principles guiding AI development, ensuring that all stakeholders are aligned in their approach. Engaging stakeholders in ethical discussions is another vital strategy. By involving diverse voices in conversations about AI use, we can ensure that a broad range of perspectives and concerns are considered. This engagement helps create AI systems that are not only technically proficient but also socially responsible.

Ending the chapter, it's clear that the ethical use of AI is a collective responsibility. As we continue to innovate and integrate AI into various aspects of life, we must remain vigilant about its implications and potential impacts. Ethical considerations are not just about preventing harm but also about maximizing AI's positive contributions to society. This sets the stage for the next chapter, where we explore how to future-proof your AI skills and adapt to an ever-evolving technological landscape.

Chapter Eight

Future-Proofing Your AI Skills

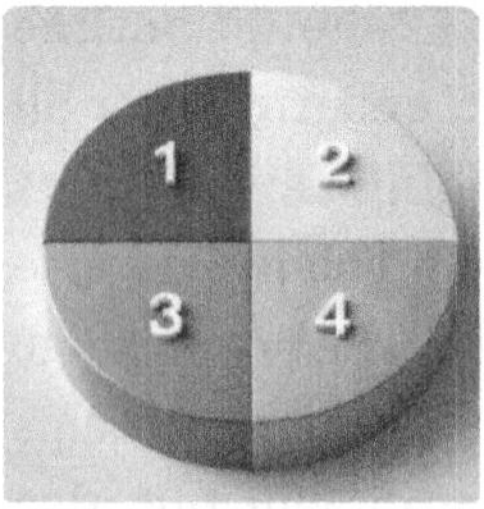

Picture this: you're at a bustling tech expo, where the air buzzes about the latest AI breakthroughs. You notice a demo of a new AI application that predicts market trends with uncanny accuracy. This scene isn't just a glimpse into the future but a reminder of how swiftly AI is evolving. Staying updated with AI trends isn't just beneficial; it's essential for leveraging AI to its fullest potential. AI technologies are advancing at breakneck speed, with new applications emerging almost daily. The opportunities are vast and ever-growing, from AI-driven medical diagnostics to AI-enhanced creative tools.

Keeping abreast of these advancements means having a finger on the pulse of innovation. Reliable sources are your allies in this quest for knowledge. Subscribing to AI-focused newsletters like "The Rundown" or "Mindstream" provides a steady stream of insights and updates. With their curated content, these newsletters ensure you stay informed without having to sift through endless articles. Industry conferences and webinars are also invaluable. They offer a platform to hear directly from experts who are shaping the future of AI. Engaging in these events can provide fresh perspectives and spark ideas to apply to your work or personal projects.

Networking plays a pivotal role in understanding AI trends. By joining AI-related professional organizations, you connect with others who share your interests and ambitions. These organizations often host events and discussions about the latest AI developments. Participating in online forums and discussions, like those on Reddit or AI Stack Exchange, expands your network and exposes you to diverse viewpoints. These interactions enhance your understanding and open doors to collaborations that can accelerate your learning.

Integrating AI learning into your daily routine might sound daunting, but it's easier than you think. Start by setting up Google Alerts for specific AI topics that interest you. This way, you receive updates directly in your inbox, keeping you informed without overwhelming you. Allocate a dedicated time each week to review AI news. It could be during your morning coffee or as part of a Friday afternoon wind-down. This habit ensures you stay current with minimal effort. Consistency is key, and by making it a regular part of your routine, you transform trend-watching from a chore into an enjoyable habit.

Interactive Element: AI News Resource List

To help you get started, here's a curated list of top AI newsletters from SOURCE 2:

- **"The Rundown"**: Offers daily insights on AI trends with over 600,000 subscribers.
- **"Mindstream"**: Provides a daily digest of AI news, perfect for quick reads.
- **"Ben's Bites"**: Known for its engaging content, available in both free and paid versions.
- **"TLDR AI"**: A concise daily newsletter focusing on the latest AI tools and innovations.

Each of these resources offers a unique perspective, ensuring you have a well-rounded understanding of the AI landscape. By tapping into these resources, you equip yourself with the knowledge needed to ask better questions and engage more effectively with AI like ChatGPT.

8.1 Preparing for the Future of AI in the Workforce

Imagine walking into a factory where machines hum quietly, performing tasks that once required hours of human labor. This isn't a scene from a distant future; it's happening now as AI reshapes industries. In manufacturing, AI automates routine tasks, boosting efficiency and reducing the potential for human error. Machines equipped with AI can predict maintenance needs, minimizing downtime and enhancing productivity. In healthcare, AI-driven innovations are transforming patient care. From AI-powered diagnostic tools that analyze medical images to personalized treatment plans based on genetic information, the potential for AI to revolutionize healthcare is immense. These changes underscore the need to prepare for a future where AI is integral to our work lives.

As AI becomes more prevalent, certain skills will be in high demand. Data literacy is crucial, as understanding and analyzing data remains a key component of AI-driven decision-making. Being able to interpret data accurately allows for better insights and more strategic choices. Statistical analysis, often the backbone of data interpretation, is another competency employers seek. Additionally, understanding AI ethics and governance becomes increasingly important. As AI systems become more autonomous, ensuring they operate within ethical guidelines is vital. This involves understanding the technology and considering its impact on society and individuals. These skills form the foundation of an AI-literate workforce capable of navigating the complexities of an AI-driven world.

To acquire these future-relevant skills, you'll need to be proactive about your learning. Enrolling in AI-focused courses and certifications is a great starting point. Many institutions offer classes ranging from basic introductions to advanced machine learning techniques. These programs provide the structured learning environment needed to grasp complex concepts. Engaging in interdisciplinary learning is also valuable. AI intersects with various fields, so understanding its applications across disciplines can offer a broader perspective. For instance, pairing AI studies with ethics or business courses can provide insights into how AI impacts different sectors. This approach not only broadens your skill set but also makes you more adaptable to changes in the job market.

The ability to adapt and engage in lifelong learning is essential in today's rapidly evolving landscape. Embracing change and innovation ensures you remain relevant as industries transform. Cultivating a culture of curiosity encourages continuous exploration and learning, driving personal and professional growth. This mindset allows you to see challenges as opportunities for development, preparing you for whatever the future holds. Building adaptability involves being open to new experiences and willing to learn from them. It's about fostering resilience and flexibility, qualities that are invaluable in a world where the only constant is change.

Textual Element: Future Skills Checklist

Consider this checklist as you plan your learning journey:

- **Data Literacy**: Understand and analyze complex datasets.
- **Statistical Analysis**: Develop the ability to interpret data accurately.
- **AI Ethics and Governance**: Learn about ethical considerations in AI use.
- **Interdisciplinary Learning**: Explore AI applications across various fields.
- **Adaptability**: Cultivate a mindset open to change and new experiences.

These skills prepare you for the future, ensuring you remain competitive and capable in an AI-driven workforce. By focusing on these areas, you position yourself as a participant in the evolving job market and as a leader ready to guide others through it.

8.2 Embracing AI as a Tool for Lifelong Learning

Imagine having a personal tutor at your fingertips, ready to guide you through any subject you wish to explore. AI has become a cornerstone in the realm of lifelong learning, offering opportunities for education that are as diverse as they are accessible. With AI, learning can be tailored to fit your personal style and pace, creating a more engaging experience. Adaptive AI platforms analyze your progress and adjust the material to match your learning style, ensuring that you grasp complex concepts at your own speed. These platforms offer personalized learning experiences, making education not only more efficient but also more enjoyable. AI-driven language learning tools, for example, provide interactive lessons that adapt to your proficiency level, helping you learn a new language quickly and confidently.

Beyond these platforms, various AI tools are designed to enhance your skills in specific areas. AI-based tutoring systems are available to assist with understanding challenging topics, offering explanations and guidance when needed. These systems can simulate real-world scenarios, allowing you to apply your knowledge practically. Interactive MOOCs (Massive Open Online Courses) have also embraced AI by incorporating AI assessments that provide immediate feedback. This instant feedback loop helps you identify areas that need improvement, encouraging continuous advancement and mastery of the subject matter. By integrating AI into these courses, learners can benefit from a more personalized and responsive educational experience, making learning more dynamic and interactive.

The advantages of AI-assisted learning are numerous, making it a valuable addition to any educational journey. Customized learning paths mean you no longer have to follow a one-size-fits-all approach.

Instead, you can focus on areas that interest you or require more attention, maximizing your learning potential. Immediate feedback allows you to track your progress effectively and adjust your learning strategies on the fly. This kind of real-time interaction motivates you to keep learning and helps you stay on track with your goals. With AI, learning becomes a continuous process, seamlessly integrated into your everyday life rather than a separate endeavor.

Consider the stories of individuals who have successfully leveraged AI for personal growth. Take the case of a marketing professional who expanded their expertise by enrolling in AI courses tailored to their field. By incorporating AI tools into their learning process, they were able to gain insights that directly impacted their work, leading to innovative marketing strategies and improved performance. Or imagine a hobbyist who used AI resources to explore new interests, like digital photography or coding. With AI's support, they transformed their passion projects into tangible skills, opening up new career opportunities and personal satisfaction. These examples highlight the transformative power of AI in lifelong learning, demonstrating how it can facilitate both professional growth and personal enrichment.

8.3 Building a Personal AI Development Plan

Crafting a personal AI development plan is a bit like mapping out a road trip. You wouldn't hit the road without knowing your destination, the stops along the way, or the best routes to take, right? A personalized AI learning roadmap operates on the same principle. It provides a structured approach to mastering AI skills, helping you establish clear goals and track milestones. By setting specific targets, you can measure your progress and identify areas needing more focus. It's crucial to pinpoint where you need to improve, whether refining your data analysis skills or understanding AI ethics more deeply. This level of clarity motivates you and gives a sense of direction, ensuring that each step brings you closer to your ultimate learning objectives.

To start building your development plan, begin with a thorough self-assessment. Identify your current skill level and knowledge gaps.

Are you familiar with basic programming, or do you need to start from scratch?

Understanding where you stand helps you select the right courses and resources tailored to your needs. Once you have a clear picture, explore the available educational options. There are numerous AI-focused courses online, ranging from beginner to advanced levels. Platforms like Coursera and edX offer a wealth of resources that cater to various learning preferences. Choose classes that align with your goals and challenge you to grow. Remember, the aim is to push your boundaries while keeping the learning process engaging and rewarding.

Executing your AI development plan requires the right tools. Digital planners and organizational apps can be invaluable in keeping you on track. Apps like Notion or Trello allow you to organize your learning schedule, set reminders, and visually track your progress. These tools help maintain focus and ensure that you're consistently working towards your goals. AI-focused educational platforms also play a significant role in executing your plan. They provide interactive learning experiences, allowing you to apply what you've learned practically. Engaging with these platforms can enhance your understanding and deepen your skills, making them an integral part of your educational journey.

It's essential to recognize that your plan is not static. Regular reviews and updates are necessary to keep it relevant and effective. Schedule periodic assessments to evaluate your progress. Are you meeting your milestones, or do you need to adjust your approach? Industry changes and new technologies might require you to shift your focus or update your goals. Being adaptable is key. Modifying your plan in response to these shifts ensures that your learning remains aligned with the latest advancements and demands. This adaptability keeps your skills sharp and prepares you for the ever-evolving landscape of AI.

Visual Element: AI Development Plan Template

Here's a simple template to help you get started:

- **Goal Setting**: Define your long-term and short-term AI learning goals.
- **Milestone Tracking**: Break down your goals into achievable milestones.
- **Skill Assessment**: Identify current skills and areas for growth.
- **Course Selection**: Choose relevant courses and resources.
- **Tool Integration**: Utilize digital planners and AI platforms.
- **Regular Review**: Schedule assessments and update goals as needed.

This template serves as a guide, helping you craft a personalized and flexible plan, ensuring you stay on the path to AI mastery.

8.4 The Role of AI in Personal and Professional Growth

Imagine you've just finished a busy workday. You're tired, but you open an AI-assisted mindfulness app. Instantly, you're guided through a calming meditation session tailored to your mood and stress levels. This is one of the ways AI can enhance personal growth. AI-driven wellness apps offer personalized experiences, helping you manage stress and improve mental health. They analyze your daily patterns, suggesting activities like breathing exercises or digital detoxes. These tools transform how you approach self-care, making it an integral part of your routine rather than an afterthought. Personalized productivity tools also play a significant role. AI can help you identify peak productivity times, organize tasks, and even suggest breaks when you're overworking. This support not only boosts efficiency but also fosters a healthier work-life balance.

On the professional front, AI is becoming a catalyst for career advancement. Consider AI-driven career coaching platforms that offer personalized guidance based on your goals and industry trends. By analyzing your career path and skills, these platforms suggest opportunities for growth and offer advice on navigating professional challenges. Automated skill assessment tools further enhance this experience. They evaluate your strengths and weaknesses, providing clear insights into areas for improvement. With these tools, you can focus your efforts on developing the skills that matter most, aligning your career trajectory with the demands of an AI-driven workforce. Such platforms empower you to take charge of your professional development, confidently making informed decisions about your career path.

Take, for instance, the story of an entrepreneur who harnessed AI to expand their business. By leveraging AI analytics, they identified new market opportunities and optimized their operations, resulting in increased revenue and growth. AI provided the data-driven insights needed to make strategic decisions, transforming challenges into opportunities. Similarly, employees are using AI to upskill and reskill, preparing for new roles within their organizations. By engaging with AI-based learning platforms, they acquire new competencies and stay ahead in a competitive job market. These stories highlight AI's transformative potential, illustrating how it can drive significant personal and professional milestones. Proactive use of AI can significantly enhance your growth strategies. Exploring AI capabilities for personal projects allows you to experiment and innovate. Whether it's using AI to refine a hobby or automate mundane tasks, the possibilities are vast. This exploration not only deepens your understanding of AI but also unlocks new avenues for personal development. Additionally, seeking AI-driven mentorship opportunities can accelerate your learning. Virtual mentors powered by AI offer personalized advice and feedback, guiding you through complex problems and supporting your growth journey. They provide a unique blend of technology and human insight, offering data-driven and empathetic guidance.

Incorporating AI into your growth strategies requires an open mind and a willingness to explore. Consider how AI can support your personal and professional goals. Reflect on the areas where AI can provide the most value, whether it's enhancing productivity, facilitating learning, or offering career guidance. By actively seeking out AI tools and resources, you empower yourself to make the most of this transformative technology. The key is to remain curious and proactive, using AI not just as a tool but as a partner in your journey towards growth and self-improvement.

8.5 Encouraging Adaptability and Continuous Improvement

In the ever-evolving landscape of AI, adaptability is not just a buzzword; it's a necessity. As technology advances at a pace that sometimes feels dizzying, being flexible and open to change becomes critical for thriving in both personal and professional arenas. The rapid technological advancements we witness today demand a mindset that is ready to pivot whenever necessary. Industries continuously shift, and expectations evolve, making it essential to stay nimble and prepared to adjust our approaches. This flexibility allows us to embrace new tools and techniques that can enhance our work and personal lives, turning potential challenges into opportunities for growth.

Cultivating a mindset of continuous improvement involves viewing challenges as learning opportunities rather than obstacles. This perspective shift can transform how we approach our goals, setting the stage for personal and professional development. By setting achievable yet challenging goals, we maintain motivation and focus, ensuring that each step we take brings us closer to our aspirations. It's about finding that sweet spot where goals are within reach but still pushing our boundaries, encouraging us to stretch and grow. This approach nurtures resilience, helping us bounce back from setbacks and learn from every experience.

Feedback plays a pivotal role in personal development, acting as a compass that guides us toward improvement and refinement. Regular feedback from peers and mentors offers valuable insights that can illuminate blind spots and highlight areas for enhancement. It's like having a coach who helps you refine your technique, ensuring you're on the right path to success. Additionally, AI tools provide a platform for self-assessment and progress tracking, offering objective feedback that can further drive growth. These tools allow us to measure our development over time, giving us the data needed to make informed adjustments and continue progressing.

Consider the stories of individuals who have thrived through adaptability. Take, for instance, a marketing professional who navigated a career shift by leveraging AI to explore new opportunities within the tech industry. By embracing new AI technologies and integrating them into their skill set, they successfully transitioned into a role that was not only rewarding but also aligned with their passions. Or think of an innovator who embraced AI advancements to propel their startup idea into reality. By staying adaptable and open to learning, they harnessed the power of AI for growth, transforming their vision into a successful enterprise. These stories underscore the value of adaptability and continuously seeking improvement, demonstrating how a flexible mindset can lead to significant achievements.

As we navigate the AI landscape, it's crucial to remain open to the idea that learning never truly ends. The ability to adapt and improve continuously is what sets apart those who merely survive from those who thrive. It's about keeping an open mind, being willing to explore new possibilities, and embracing the changes that come our way. By fostering this mindset, we equip ourselves with the tools needed to succeed in a constantly changing world.

Quiz: Have you progressed through the Four learning and skill acquisition stages?

Conclusion

A Journey of Empowerment and Innovation

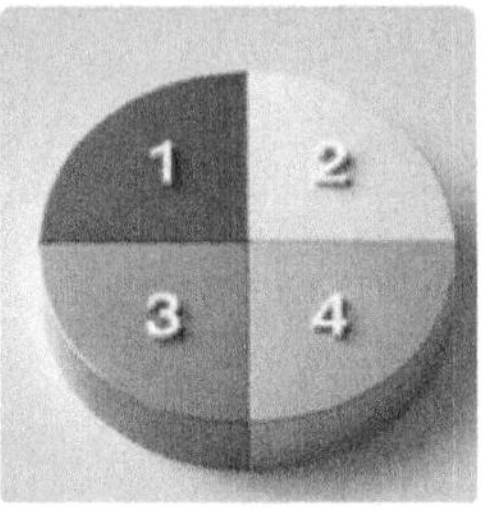

As we reach the conclusion of ***Unlocking ChatGPT for Beginners***, it's clear that your journey with conversational AI has only just begun. This book has served as a guide to exploring the vast capabilities of ChatGPT, offering foundational knowledge, practical strategies, and forward-looking insights to help you unlock the potential of this transformative technology.

Reflecting on Key Takeaways

Throughout this book, we've delved into the mechanics of ChatGPT and its application across diverse areas of personal and professional life. From mastering the art of crafting effective questions to overcoming barriers and leveraging advanced prompt strategies, you've gained tools to maximize the value of AI in real-world contexts. Each chapter has provided stepping stones for fostering confidence, curiosity, and adaptability in your AI interactions.

- **Foundations of ChatGPT** introduced you to the core principles and capabilities of this groundbreaking tool, setting realistic expectations while empowering you to navigate its interface with ease.
- **Crafting Effective Questions** emphasized the importance of clarity, context, and iteration in generating meaningful and relevant responses.

- **Overcoming Barriers** addressed common challenges in AI adoption, offering practical strategies to build trust, refine outputs, and save time.
- **Real-World Applications** showcased the versatility of ChatGPT in creative, educational, and professional settings, demonstrating how it can elevate your productivity and problem-solving abilities.
- **Interactive Learning and Continuous Improvement** encouraged you to embrace a mindset of experimentation and feedback to deepen your expertise over time.
- **Advanced Prompt Strategies** and **Building AI Literacy** equipped you with techniques to unlock more sophisticated use cases while fostering a nuanced understanding of AI technologies.
- Finally, **Future-Proofing Your AI Skills** inspired you to view AI as a lifelong partner in growth and adaptability, preparing you for an innovation-defined future.

Looking Ahead: The Evolving Role of AI

The rapid evolution of AI technologies means that new opportunities, applications, and challenges will continue to emerge. As you integrate ChatGPT into your daily life, remember that adaptability is the key to staying ahead. By remaining curious, informed, and open to learning, you'll keep pace with advancements and help shape how these tools are used in your personal and professional circles.

AI is not just a tool—it's a collaborator. The more you invest in understanding its strengths and limitations, the more effective and rewarding your interactions will become. As you continue your exploration, remember to approach AI with enthusiasm and responsibility, recognizing its ethical implications and transformative potential.

Your Next Steps

Now that you've completed this book consider the following actions to solidify your knowledge and skills:

1. **Practice Regularly:** Experiment with ChatGPT in various contexts to reinforce your learning and discover new use cases.

2. **Engage with the AI Community:** Join forums, workshops, or online groups to exchange ideas, share insights, and stay updated on the latest developments.

3. **Embrace Lifelong Learning:** Make AI literacy a cornerstone of your personal and professional development plan, leveraging it as a tool for growth and innovation.

4. **Advocate for Ethical AI Use:** Promote responsible AI adoption in your community, emphasizing transparency, fairness, and inclusivity.

A Final Thought

As you finish this book, remember that your journey with ChatGPT doesn't end here. It continues with every conversation you start, every problem you solve, and every insight you gain through the power of AI. **The future belongs to those who embrace change, pursue knowledge, and harness technology to create meaningful impact.**

Move forward with confidence, creativity, and curiosity—ready to explore AI's limitless opportunities. The world of ChatGPT is now yours to discover.

Don't forget to check out the bonus section for additional insights.

Wishing you great success as you unlock new possibilities with ChatGPT and beyond!

Bonus Content

The Five W's

The Five W's framework is a powerful tool for crafting highly effective responses to your questions. By using **who, what, where, when,** and **why**, you can create well-structured and clear prompts that help define your needs, enabling ChatGPT to provide more relevant and precise answers. Here's how each aspect contributes to this process:

1. **Who**: Define the audience, participants, or key individuals.
   - Example: *Who are the primary stakeholders in senior housing appraisals?*
2. **What**: Specify the main topic, goal, or problem.
   - Example: *What are the latest trends affecting nursing home appraisals in the Midwest?*
3. **Where**: Add geographic or contextual relevance.
   - Example: *Where are the best locations for developing memory care facilities in the U.S. Midwest?*
4. **When**: Include time-related context if needed.
   - Example: *When do assisted living facilities typically see peak occupancy rates?*

5. **Why**: Explain the purpose of your inquiry.
    - Example: *Why do appraisers adjust for location in valuations of senior housing facilities?*

Applying This Framework to Build a ChatGPT Prompt

Combine these questions to form a detailed and contextually rich prompt. For example:

- "**Who** are the primary stakeholders, **what** are the factors influencing them, and **why** are location adjustments critical in appraising nursing homes in the Midwest today?"

This process ensures your prompt is targeted and comprehensive, allowing for a detailed and tailored AI response. Here are several examples of a prompt that uses the 5W's framework effectively:

Prompt 1:

- "Imagine you're helping a commercial real estate appraiser specializing in senior housing. **Who** are the key stakeholders involved in the appraisal process? **What** factors should be considered when valuing assisted living and memory care facilities? **Where** are the most significant regional trends influencing senior housing in the Midwest? **When** should appraisers anticipate seasonal occupancy changes, and **why** do these fluctuations impact property valuations?"

This prompt incorporates **who, what, where, when, and why** to provide clear context, guide the AI's response, and ensure relevant and detailed output.

Prompt 2:

- "Using the 5W's framework, brainstorm a compelling title for a book about the evolution of senior housing. **Who** are the key players shaping the industry? **What** challenges and innovations are transforming the field? **Where** are the most notable developments taking place? **When** did the pivotal shifts occur, and **why** is this topic relevant for the future of aging populations?"

This prompt also incorporates **who, what, where, when,** and **why** to provide clear context, guide the AI's response, and ensure relevant and detailed output.

Some prompts may be helpful for repetitive tasks and worth saving for future use. However, the foundation of a compelling question lies in **clarity, context**, and **specificity**—key elements that can be easily achieved using the 5W's framework.

The Prompt Library

Do you need a library of 200 ChatGPT prompts? The answer is no. By honing your questioning skills and leveraging the 5Ws, such as **who, what, where, when,** and **why**, you can create tailored prompts that address nearly any situation you encounter, both personally and professionally.

The Four Stages Of Competence

The Four Stages of Competence model, also known as the Conscious Competence Ladder, is attributed to Noel Burch, who developed it in the 1970s while working for Gordon Training International. This model was initially created to help trainers understand and guide the process of learning and skill acquisition.

The model has since been widely adopted in various fields, including education, corporate training, and personal development, due to its simplicity and applicability to diverse learning processes.

We introduced this concept in the introduction to this book. We will now give you a more in-depth look at how the Four stages can be used in your daily life. Mapping the **four stages of competency** into the process of **learning** can provide insight into how individuals progress when acquiring new knowledge or skills:

1. Unconscious Incompetence

- **Relation to Learning:**
  - At this stage, the learner is unaware of their lack of knowledge or skill. They might not understand the importance of learning or even recognize the existence of the subject.
- **Example in Learning:**
  - A student who doesn't know how to solve algebraic equations is unaware of their relevance or difficulty.
  - Someone who believes they can master a foreign language simply by watching movies without formal study.

2. Conscious Incompetence

- **Relation to Learning:**
  - The learner becomes aware of their knowledge gap. They realize they don't know or understand something, which can often be a humbling experience. This stage is critical for motivation as it triggers the desire to learn.
- **Example in Learning:**
  - The student struggles with algebraic equations in class and realizes they don't understand the steps involved.
  - A person trying to learn a new language struggles with

pronunciation or grammar, realizing how much effort is required.

3. Conscious Competence

- **Relation to Learning:**
    - The learner starts acquiring knowledge and practicing the skill. They can perform tasks correctly, but it requires focus, effort, and conscious thought. Mistakes may still occur, but progress is evident.
- **Example in Learning:**
    - The student practices algebra step by step, frequently checking their work to ensure they follow the rules of solving equations.
    - The language learner can form basic sentences and communicate in simple terms but must think carefully about grammar and vocabulary before speaking.

4. Unconscious Competence

- **Relation to Learning:**
    - The learner has mastered the subject or skill. It becomes second nature, requiring little conscious effort. They can now teach or apply their knowledge seamlessly.
- **Example in Learning:**
    - The student solves algebraic equations automatically and confidently, even explaining the process to others.
    - The language learner speaks fluently and naturally, constructing sentences and conversing without needing to think about grammar or vocabulary.

Key Takeaways:

- **Learning is a progression:** Moving through the stages involves practice, feedback, and time.
- **Teaching strategies differ by stage:**
    - In **stage 1**, the focus should be on awareness and motivation.
    - In **stage 2**, educators should offer structured guidance and resources.
    - In **stage 3**, practice and repetition are essential, with frequent feedback.
    - In **stage 4**, the learner can apply and even teach the knowledge.

By understanding these stages, educators and **learners like you** can tailor their approaches to maximize learning outcomes.

You may not realize that you are going through the four stages every time you learn something new.

1. **Unconscious Incompetence**
    - **Example:** Someone who doesn't know how to drive but isn't aware of the complexities involved in learning to drive.
2. **Conscious Incompetence**
    - **Example:** A person starts driving lessons and realizes how much there is to learn.
3. **Conscious Competence**
    - **Example:** A new driver can drive a car but must focus on every action, such as checking mirrors and shifting gears.

4. **Unconscious Competence**

    - **Example:** An experienced driver drives intuitively, without actively thinking about the mechanics of driving.

One of our earliest memories might be learning how to ride a bike. Eager to give it a try, we quickly realized just how challenging balancing could be. Training wheels became our trusty allies, helping us gain both stability and confidence. After countless tumbles, scrapes, and band-aids, we eventually mastered the skill, riding freely without sacrificing our clothes—or more skin—to the unforgiving pavement.

Whether it is a bike, a car, or another life experience, we all go through this process. Novice or expert, we all need to go through these learning stages to be proficient in all aspects of life.

Quiz: What stage are you in now?

I hope your answer was Stage 1 and you are on the path to more learning in other areas of interest. Don't forget, we all start in stage 1.

Book Review

Keeping the Conversation Flowing

You've now unlocked the tools to harness the power of ChatGPT and make it part of your daily life. But the journey doesn't stop here—it's your turn to help others discover the same possibilities!

By sharing your honest review of this book on Amazon, you'll guide others curious about ChatGPT but unsure where to begin. Your insights could be the encouragement they need to take their first step.

Your review could help someone:

- Learn how ChatGPT can simplify their day-to-day tasks.
- Gain confidence in exploring AI for the first time.
- Find creative ways to unlock their potential with ChatGPT.

Thank you for being part of this journey. When we share what we've learned, we keep the spirit of discovery alive—and you're helping to make that happen.

Click here to leave your review on Amazon.

Thanks,

C.P. Bailey

References

- *ChatGPT Architecture Explained. - Medium* https://medium.com/@sreedevr/chatgpt-architecture-explained-7fc0cbb7426c#:~:text=Overall%2C%20the%20architecture%20of%20ChatGPT,and%20a%20dialogue%20generation%20head.

- *From Eliza to ChatGPT: the 60-year history of chatbots* https://www.theverge.com/24054603/chatbot-chatgpt-eliza-history-ai-assistants-video

- *11 Real-Life Examples of NLP in Action* https://levity.ai/blog/11-nlp-real-life-examples

- *How to Use ChatGPT: The Full Guide for Beginners* https://careerfoundry.com/en/blog/digital-marketing/how-to-use-chatgpt/

- *Prompt Engineering Best Practices: Tips, Tricks, and Tools* https://www.digitalocean.com/resources/articles/prompt-engineering-best-practices

- *5 Common Generative AI Prompt Writing Mistakes (And ...* https://www.forbes.com/sites/bernardmarr/2024/10/01/5-common-generative-ai-prompt-writing-mistakes-and-how-to-fix-them/

- *Mastering Generative AI Interactions: A Guide to In-Context ...* https://www.linkedin.com/pulse/mastering-generative-ai-interactions-guide-in-context-pradeep-menon

- *Unlock AI's Potential Through Iterative Prompting* https://medium.com/@nathanarodgers/unlock-ais-potential-through-iterative-prompting-974ab5460525
- *50 AI Terms Every Beginner Should Know | TELUS Digital* https://www.telusdigital.com/insights/ai-data/article/50-beginner-ai-terms-you-should-know
- *The Psychological Fears Associated With AI* https://www.psychologytoday.com/us/blog/beyond-stress-and-burnout/202307/the-psychological-fears-associated-with-ai
- *AI Strategies Series: Inconsistent and Outdated Responses* https://insight.factset.com/ai-strategies-series-inconsistent-and-outdated-responses
- *Collaborative Intelligence: Humans and AI Are Joining ...* https://hbr.org/2018/07/collaborative-intelligence-humans-and-ai-are-joining-forces
- *How to Automate ChatGPT to Become a Productivity Guru* https://airbyte.com/data-engineering-resources/automate-chatgpt
- *Our Artificial Intelligence Client Success Stories | BCG* https://www.bcg.com/capabilities/artificial-intelligence/client-success
- *6 Tips for Using ChatGPT to Brainstorm Better* https://northwest.education/insights/career-growth/6-tips-for-using-chatgpt-to-brainstorm-better/
- *Innovative Techniques for Implementing ChatGPT in Remote ...* https://cluelabs.com/blog/innovative-techniques-for-implementing-chatgpt-in-remote-learning/
- *AI Prompt Engineering Techniques: The Power Of Crafting ...* https://slash.co/articles/ai-prompt-engineering-techniques

- *The AI Feedback Loop - Continuous Improvement in ...* https://www.linkedin.com/pulse/ai-feedback-loop-continuous-improvement-people-leonard-langsdorf-sl4ve

- *The Best AI Communities Online* https://aiproducthive.com/index.php/2024/05/27/the-best-ai-communities-online/

- *Best AI LMS Tools For Training And Education In 2024* https://elearningindustry.com/best-ai-tools-for-training-and-education-top-lms-platforms

- *Creating complex prompts for ChatGPT: Advanced ...* https://medium.com/@quanticascience/creating-complex-prompts-for-chatgpt-advanced-techniques-for-more-accurate-results-37bb9e76172c

- *How AI Can Help Leaders Make Better Decisions Under ...* https://hbr.org/2023/10/how-ai-can-help-leaders-make-better-decisions-under-pressure

- *Prompts that Work: Enhancing AI Interactions with Sequenced ...* https://susanmernit.medium.com/prompts-that-work-enhancing-ai-interactions-with-sequenced-multi-step-prompts-b55af4f0a2e6

- *AI Personalization Examples and Challenges* https://www.bloomreach.com/en/blog/ai-personalization-5-examples-business-challenges

- *AI Literacy: A Framework to Understand, Evaluate, and Use ...* https://digitalpromise.org/2024/06/18/ai-literacy-a-framework-to-understand-evaluate-and-use-emerging-technology/

- *Exploring AI Tech Stack: Essential Components and ...* https://www.moontechnolabs.com/blog/ai-tech-stack/

- *The Role of Artificial Intelligence in Digital Transformation* https://www.ptc.com/en/blogs/corporate/artificial-intelligen

ce-digital-transformation

- *Ethical concerns mount as AI takes bigger decision-making ...* https://news.harvard.edu/gazette/story/2020/10/ethical-concerns-mount-as-ai-takes-bigger-decision-making-role/

- *AI Index Report 2024 - Stanford University* https://aiindex.stanford.edu/report/

- *Top 17 AI Newsletters for 2024* https://explodingtopics.com/blog/ai-newsletters

- *The top 10 employee skills needed for artificial intelligence* https://www.multiverse.io/en-GB/blog/the-top-10-employee-skills-needed-for-artificial-intelligence

- *ChatGPT and Lifelong Learning -* https://www.missiontolearn.com/chatgpt-and-lifelong-learning/

Made in the USA
Monee, IL
04 November 2026

67381926R00075